AF527095

FUNDAMENTALS OF SEDIMENTARY ROCKS

By

N.W. Gokhale

M.Sc., Ph.D. (Budapest), F.M.S. (Mysore), F.I.S.E.S. (Pune)

Professor of Geology, Department of Studies in Geology,
Karnataka University, Dharwad (Karnataka)

CBS

CBS Publishers & Distributors Pvt. Ltd.

New Delhi • Bengaluru • Chennai • Kochi • Kolkata • Mumbai
Hyderabad • Nagpur • Patna • Pune • Vijayawada

ISBN: 81-239-0621-8

First Edition: 1998
Reprint: 2003, 2004, 2009, 2011, 2013, 2017

Copyright © Author & Publisher

All rights reserved. No part of this book may be reproduced or transmitted in any form or by any means, electronic or mechanical, including photocopying, recording, or any information storage and retrieval system without permission, in writing, from the publisher.

Published by **Satish Kumar Jain** and produced by **Varun Jain** for
CBS Publishers & Distributors Pvt. Ltd.,
4819/XI Prahlad Street, 24 Ansari Road, Daryaganj, New Delhi - 110002
delhi@cbspd.com, cbspubs@airtelmail.in • www.cbspd.com
Ph.: 23289259, 23266861, 23266867 • Fax: 011-23243014

Corporate Office: 204 FIE, Industrial Area, Patparganj, Delhi - 110 092
Ph: 49344934 • Fax: 011-49344935
E-mail: publishing@cbspd.com • publicity@cbspd.com

Branches:
- ***Bengaluru:*** 2975, 17th Cross, K.R. Road, Bansankari 2nd Stage, Bengaluru - 70 • Ph: +91-80-26771678/79 • Fax: +91-80-26771680 E-mail: cbsbng@gmail.com, bangalore@cbspd.com
- ***Chennai:*** No. 7, Subbaraya Street, Shenoy Nagar, Chennai - 600030 Ph: +91-44-26681266, 26680620 • Fax: +91-44-42032115 E-mail: chennai@cbspd.com
- ***Kochi:*** Ashana House, 39/1904, A.M. Thomas Road, Valanjambalam, Ernakulum, Kochi • Ph: +91-484-4059061-65 Fax: +91-484-4059065 • E-mail: cochin@cbspd.com
- ***Kolkata:*** 6-B, Ground Floor, Rameshwar Shaw Road, Kolkata - 700014 Ph: +91-33-22891126/7/8 • E-mail: kolkata@cbspd.com
- ***Mumbai:*** 83-C, Dr. E. Moses Road, Worli, Mumbai - 400018 Ph: +91-9833017933, 022-24902340/41 • E-mail: mumbai@cbspd.com

Representatives:

- Hyderabad: 0-9885175004
- Patna: 0-9334159340
- Vijayawada: 0-9000660880
- Nagpur: 0-9021734563
- Pune: 0-9623451994

Printed at:
J.S. Offset Printers, Delhi (India)

Foreword

At a time when countries around the world are looking for more natural resources in their race to compete with the others to provide a welfare society, geology invariably becomes an important discipline. Being a subject that deals with the matter beneath the earth, various dimensions of understanding the formation of earth would lead us to understand the importance of layers of earth which in turn stresses on the study of fundamentals of sedimentary rocks.

Geology is essentially a field science. A field geologist, as he is often recognised, concentrates on this kind of study which is rather a tedious one. The amount of field work involved and the mental strain that he has to bear, makes this job of a field geologist a tiresome work. This book entitled "Fundamentals of Sedimentary Rocks" authored by Dr. N.W. Gokhale helps us to understand and share the knowledge of the endeavours of such a field geologist, who has had a long innings as a "geologist in the field", in the true sense of the term. This book, as the title itself suggests, deals with the fundamental aspects of sedimentary rocks covering among other things the genesis, structure, texture and analysis of sedimentary rocks. The book has seven chapters covering various aspects of sedimentary rocks. A large number of field photographs documenting structural features, particularly from Kaladgi basin, make this book truly fundamental and interesting, helping the new entrants in this area to understand the importance of this study.

Prof. Gokhale, a dedicated teacher with nearly three decades of experience in teachings at the Post-Graduate Department of Geology, Karnatak University, Dharwad, has made a sincere effort at presenting the facts related to the study of sedimentary rocks in a systematic way. Chapter 1 deals with the importance of sedimentary rocks and their classification, taking into consideration the different bases.

A chapter on the origin of the rocks and the various stages of its formation, is lucidly presented and this helps a reader to understand the details of formation of rocks and its presence under the surface of the earth. The author has taken care to cover the information on both the sedimentary rocks and the sedimentary analysis. This will help a serious student of geology, to locate information relating to sedimentary rocks in full in this book, which is otherwise scattered. This book presents new types of diagrammatic presentation such as superimposition of graphs etc. to help to comment on the grain size characteristics of sedimentary rocks in a better manner. A trivariate diagram based on "Mean Size, Skewness and Standard Deviation" seems to be first of its kind. This helps a reader to demarcate the fields of "River and Beach" environments rather more precisely. A three-dimensional diagram has also been drawn to depict the notion of **two surfaces** controlling the process of sedimentation, namely, Sea level and Sediment-Water Surface. These additions, thus, make reading more informative and Prof. Gokhale's attempt in this direction needs to be plauded.

I congratulate Prof. Gokhale for having taken up this useful work. He, no doubt, deserves the accolades both from the students and the teachers alike. I am sure this book will be a welcome addition to the literature in Geology.

Dr. A.M. Pathan
Vice-Chancellor, Karnatak University,
Pavate Nagar, Dharwad - 580 003,
Karnataka, India

Preface

In this book, I have endeavoured to elaborate the essential aspects of the sedimentary rocks. Geological studies commence with field work, therefore the text of this book commences with the field features of these rocks. This is followed by the structural, the textural, the petrographical and the petrochemical characteristics of the sedimentary rocks. The evolution of these rocks has been presented in six stages in the form of a flow-sheet, bringing forth the interdependence and the inter-relations between the several stages.

Geological accounts require documentation in the form of photographs. In this context I deem it my good fortune in getting excellent field photographs and photomicrographs from Doctors N.N. Gothe (Professor and Head of Department of Mineral Processing and Beneficiation, P.G. Centre of Gulbarga University, Krishnadevarayanagar, Sandur, Bellary district, Karnataka), V.B. Koppad (Professor and Head of Geology Department, H.K. Society's P.D.A. Engineering College, Gulbarga, Karnataka), S.C. Puranik (Professor of Geology, Karnatak University, Dharwad), H.D. Desai (Indian Bureau of Mines, Nagpur), M.J. Chandrasekhara Gowda (Professor of Geology, Department of Geology, Mysore University, Mysore, Karnataka), G.V. Hegde (Assistant Geologist, Karnataka State Department of Mines and Geology, Bangalore, Karnataka), G.S. Pujar (Lecturer in Geology, Department of Geology, Karnatak Science College, Dharwad, Karnataka), K. Bhimsen (Geologist, Mentone, Victoria, Australia), and A.H. Kouhsari (Professor of Geology, Department of Mining Engineering, Yazd, Iran). I am thankful to all of them for sparing their precious photographs to me. These photographs have indeed enhanced the quality of the text of this book.

The Doctoral Theses of G.V. Hegde, G.S. Pujar and K. Bhimsen were of immense help to me while drafting some parts of this book. I place on record my sincere thanks to them for sparing the said theses to me. I am grateful to Dr. P.S. Agarkar (Retired Professor of Geology, Science College, Jabalpur, M.P.) for his critical review of the text, and useful suggestions. Some of the field photos are indeed unique; like the colour banded quartzarenites, and the colour spotted quartzarenites. Some novel diagrams have been presented in this book. The chapter dealing with the "Academic and Economic Importance" of the sedimentary rocks is a special feature of this book.

Requirement of the researchers namely an account of the "Sedimentary Analysis" is not lost sight of! Any student of geology who is desirous of acquiring knowledge about the sedimentary rocks, will find this book very useful.

Last but not the least, I place on record my heartfelt gratefulness to Dr. A.M. Pathan, Vice-Chancellor of the Karnatak University, Dharwad, for kindly penning the foreword to this book.

N.W. Gokhale

Contents

Frontispiece. Exposure of pink coloured quartzarenites 1.5 km N 68° W of Huli village, Saundatti taluka, Belgaum district, Karnataka state. Note the piling up the beds one over another. Current bedding is observable in the central part of the exposure. The beds are nearly 2 metres thick. View looking S 80° W. *Courtesy :* Dr. G.S. Pujar.

CHAPTER 1

Distinctive Features of Sedimentary Rocks

INTRODUCTION

The crust of the earth is made up of three varieties of rocks, namely the igneous, the sedimentary and the metamorphic. The sedimentary rocks possess certain features that are not associated with the other two varieties of rocks. The distinctive features can be categorised into

1. Field features,
2. Petrographic characters,
3. Textural and structural features, and
4. Outcrop patterns and shapes.

It is the mode of origin that imparts certain special features to the sedimentary rocks. To ascertain the exact mode of origin of the sedimentary rocks is many times not possible, but the rest of the features could be observed in these rocks without entertaining any doubts in the mind. The very word "sedimentary" already conveys the fact that sediments are necessary in the development of these rocks. The word "sediment" in turn points to the role of rivers, because, it is a common knowledge that, in the main, the sediments are transported by the rivers, besides by the other geological agents. The rivers and the wind transport the sediments produced by the weathering of the pre-existing crustal rocks, and such sediments ultimately find resort in the "inland basins", or "the seas and the oceans". The sediments are required to be converted into hard and compact rocks which are then termed as the sedimentary rocks. Some of the distinctive features of these rocks have been elaborated in the following pages.

FIELD FEATURES

Sedimentary rocks have an unique mode of occurrence in the crust of the earth. As these are bedded formations, the different beds are seen to be piled one over another (Figs. 1.1a, b, c and frontispiece photo). The geomorphic expression of these rocks is also unique. Flat topped hills are normally produced, if the sediments were not disturbed during the upheaval of the basin. The flat topped hills of Tirupati, Andhra Pradesh (Fig. 1.2), the hills of Godachimalaki, Gokak taluka, Belgaum district, Karnataka (Fig. 1.3) are good examples, because horizontal beds of sandstones compose the said hills, and these rocks in turn have controlled the shapes of the hills. However during the upheaval of the basin, the rocks might get tilted, and the geomorphic expression of such rocks is the "dip slope and escarpment scenery" and the "hogback scenery". Sandstones exposed in the vicinity of Gokak town, Belgaum

Fig. 1.1a. Beds of gray coloured quartzarenites piled one over another. The beds are dipping 78° due S 74° W. The rocks are exposed at the Kolchi weir on the Malaprabha river, Ramdurg taluka, Belgaum district, Karnataka state. View looking north. *Courtesy :* Dr. G.V. Hegde.

Fig. 1.1b. Light pink coloured sandstone beds exposed in the vicinity of Godachimalaki, Gokak taluka, Belgaum district, Karnataka. Note the variation of thickness of beds, and the piling up of beds one over another. The beds are almost horizontal in attitude. *Courtesy :* Dr. G.S. Pujar.

Fig. 1.1c. Cream coloured quartzarenites exposed in the Hiranyakeshi river. The beds display a variation of thickness from 50 cm (thick) and 6 cm (thin), and piling up of one bed over another. The beds possess a low dip of 10° due N 10° W. The exposure is located S 10° E of Ramtirth temple, Ajra, Kolhapur district, Maharashtra. *Courtesy :* Dr. K. Bhimsen.

Fig. 1.2. Flat topped hills of Tirupati, Andhra Pradesh. Note the horizontality of the beds of sandstones, which in turn have given rise to the flat top of the hills. The bottom-most bed marks the plane of "Eparchaean unconformity", it starts right at the conspicuous break of slope from the vertical (cliff) to relatively gentler (50-60°) slope. *Courtesy :* Dr. G.S. Pujar.

Fig. 1.3. Typical flat topped hills produced by the horizontally reposing sandstones exposed in the southwestern part of the Kaladgi basin. Locality is the Godachimalaki village, Gokak taluka, Belgaum district, Karnataka state. *Courtesy :* Dr. G.S. Pujar.

district, Karnataka (Fig. 1.4), and sandstones and the other associated rocks constituting the Vindhyan mountains, display the "dip slope and escarpment" topography. The point to be noted is that unless the beds (rocks) possess high inclination, the sedimentary rocks cannot give rise to the ridge-like hills which are usually produced by the basic igneous dykes. Caution is to be exercised in respect of the Deccan Traps of India which have given rise to the flat-topped hills (plateau scenery) only because the flows are almost horizontal in attitude. Therefore from the geomorphic point of view it can be argued that as a first approximation, the sedimentary rocks (and not the igneous or the metamorphic rocks) be reasonably expected, if flat-topped hills be noticed in any area. Of course, the feebly or strongly metamorphosed equivalents of the sedimentary rocks might also display such a topography. But then there are other petrological characters that help to make the distinction between the said two varieties of rocks. Limestones display pitted appearance owing to chemical weathering. Such rocks alone produce the stalagtites and the stalagmites. The topography in general is described by the term "karst". Being a softer rock, valleys are initiated on it, rather than on the associated relatively harder rocks like the sandstone, conglomerate etc.

Fig. 1.4. Beds of sandstones exposed in the vicinity of Gokak town, Belgaum district, Karnataka, displaying "dip-slope escarpment" scenery (background part of the photo). The hanging bridge is on the river Ghataprabha, which has developed a 55-metre deep waterfall in the vicinity of Gokak town. *Courtesy :* Dr. G.S. Pujar.

The presence of layers (called as beds) is another unique feature owing to which these rocks produce "tabular bodies". The individual layers might differ in colour, grain size or due to a combination thereof. Igneous rocks do not possess layers excepting the lava flows. The metamorphosed sedimentary rocks (which are technically called as meta-rocks) retain the pre-existing bedded nature and the other primary sedimentary structures (Figs. 1.5a, b, c). Banded hematite quartzite is an unambiguous example of it, which previously was a layered siliceous ferruginous sedimentary rock (Fig. 1.5a). Blocks of rocks developed through combination of joint

Fig. 1.5a. Hand-specimen of banded hematite quartzite (3/4th natural size) displaying a great variation in thicknesses of the ferruginous and the siliceous layers (bands). Multitudinous bands are developed within a thickness of a mere 12 cm. Such a structure is designated as a "Ghost or a Relict" structure. The rock sample is from the Jambunath hills, N.E.B. range, Bellary district, Karnataka state. *Courtesy :* Dr. H.D. Desai.

Fig. 1.5b. Hand-specimen of a banded hematite quartzite showing "ripple marks". These are typical oscillation ripple marks. This structure belongs to the "Ghost/Relict" category. The specimen is from the famous auriferous Kappatgudda hill ranges, locality being Nagavi village, Dharwad district, Karnataka state. *Courtesy :* Dr. S.C. Puranik.

Fig. 1.5c. Field exposure of "robust oscillation ripple marks" developed in banded iron quartzites encountered 6 km N 20° W of Vadrahalli village, Bellary district, Karnataka state. This structure belongs to "Ghost/Relict" category. *Courtesy* : Dr. H.D. Desai.

planes (called as joint-blocks) in the igneous or the metamorphic rocks seldom attain the dimensions of the sedimentary beds. Therefore from the layered or the bedded nature, existence of large sized joint-blocks, the presence of the sedimentary rocks can be reasonably inferred.

Besides the existence of bedding planes, the sedimentary rocks possess current bedding (Figs. 1.6a, b), graded bedding (Fig. 1.7) and ripple marks (Figs. 1.8a, b, c), which are considered as the most typical structural features of these rocks. Animal and plant fossils are characteristically and singularly associated with these rocks (Figs. 1.9a, b). Thus a careful study of the field features enables the geologists to infer with a fair degree of accuracy, the presence or otherwise of the sedimentary rocks in the area selected for investigation.

PETROGRAPHIC CHARACTERS

Unlike the igneous and the metamorphic rocks, not many varieties are met with, within the sedimentary rocks. Further, almost every variety is "monomineralic" in composition. Thus in a sandstone, almost the entire rock is made up of grains of quartz; clay minerals (kaolinite, illite, montmorillonite etc.) compose a shale, and so on. Therefore a monomineralic composition is a clear pointer to the existence of a sedimentary rock, barring some igneous rocks (like the dunite, the anorthosite, the hornblendite etc.) which are also made up of only one or two minerals. But these minerals are totally different from those occurring in the sedimentary rocks. Alongside with monomineralism, the shapes of the grains of the sedimentary rocks, are very diagnostic ones. The grains are rounded to subrounded in the majority of cases.

Fig. 1.6a. Vertical section of current bedded structure developed in the quartzarenites exposed 1.[illegible] km N 70° E of Nilgund, Badami taluka, Bijapur district, Karnataka state. Six to seven sets of current bedded units are observable. The rocks are dipping away from the observer at an angle of 20°. *Courtesy :* Dr. G.V. Hedge.

Fig. 1.6b. Field photo of current bedding developed in sandstones. Note the development of several sets of current bedding. Also note the gentle angle of repose at the lower part, whereas abrupt termination (high angle of repose) at the upper part of the "off-set beds". Locality : Amingarh town, Bijapur district, Karnataka state. *Courtesy :* Dr. S.C. Puranik.

Fig. 1.7. Photomicrograph showing graded bedding developed in the specimen of quartzarenite collected from the southeastern part of the Kaladgi basin, Karnataka state. Crossed nicols ×3.5. *Courtesy* : Dr. G.V. Hegde.

Euhedral grains or those having sutured borders, are nearly absent. Further, the grains are held together by a cementing material. Therefore, rounded to subrounded grains set in a cement, is the most salient feature of the sedimentary rocks. One another unique feature is the presence of a variety of heavy minerals (rutile, ilmenite, tourmaline, sphene etc.) in these rocks. However the heavy minerals are required to be identified by adopting special techniques, because these cannot be identified through the routine thin section microscopic examination.

TEXTURAL AND STRUCTURAL FEATURES

By far the most salient or distinctive feature of the sedimentary rocks is the presence of binding material (called as matrix or cement) for the grains. Neither the igneous nor the metamorphic rocks display such a feature. The grains are held together in a variety of matrix. This imparts a special textural appearance to the sedimentary rocks. A mention has already been made to the layered or the bedded nature of these rocks. This, in fact, is a textural character. The bedded and the banded nature is due to the sorting of grains (sediments) according to the different sizes during the process of sedimentation. At times the layers are so thin that it results in the development of a laminated structure. However, care should be taken not to mistake "rock-cleavage" to lamination, because at first sight, these two structures almost look

Fig. 1.8a. Hand-specimen showing oscillation ripples developed in quartzarenite, the specimen being collected 3 km S 70° E of Belavalkop village, Badami taluka, Bijapur district, Karnataka state. Note the concavity of the ripples facing upwards. *Courtesy :* Dr. G.V. Hedge.

Fig. 1.8b. Excellently developed current ripples in the sandstones of Navapur village, Gokak taluka, Belgaum district, Karnataka state. *Courtesy :* Department of Geology, Karnataka Science College, Dharwad.

Fig. 1.8c. Unusually large sized current ripples developed in the quartzarenites exposed on the west bank of the Hiranyakeshi river at the Ramtirth temple, Ajra, Kolhapur district, Maharashtra state. The distance between two consecutive crests is around 34 cm with a height of 6 cm. The rocks trend N 70° W - S 70° E with a dip of 10° due N 20° E. These ripples belong to the category of "Giant ripples". *Courtesy :* Dr. K. Bhimsen.

Fig. 1.9a. Field exposure of horizontal colonies of "Giant" algal stromatolites developed in variegated limestones. Locality : Alagundi town, Bijapur district, Karnataka state. *Courtesy :* Dr. M.U. Chandrasekhara Gowda.

Fig. 1.9b. Exposure of vertical columns of algal stromatolites developed in cherty limestones. Locality . Jambigi town, Bijapur district, Karnataka state. *Courtesy* : Dr. M.J. Chandrasekhara Gowda.

alike. The other distinctive features like the current bedding and the graded bedding have already been noted. There are still others like the rain prints (Figs. 1.10a, b), the mud or sun cracks, the worm burrows and so on, which will be described in a separate chapter. Some of the textures/structures are so distinctive and hence typical, that their very presence stamps the rock as belonging to the sedimentary category. These above noted structures are very easily and readily observable during the field studies.

OUTCROP PATTERNS AND SHAPES

Though this is a field feature, it however cannot be immediately appreciated, until the rocks cropping out in the area are accurately and completely mapped. The outcrop pattern and the shape of the sedimentary rocks has a direct bearing on the mode of origin. The sediments are laid in a basin, and therefore, the shape and the size of the outcrops reflect that of the basin. The sedimentary basins are observed to be more long than broad, and these possess certain uniform depth. The sediments and the rocks formed therefrom, therefore imitate the characters of the basin. Basin is a regular body, and the sediments laid in it therefore, acquire a tabular form. It thus becomes possible to predict the length (extent of the rocks), the breadth and the depth (thickness) of these rocks. Some examples have been cited to illustrate these features. The exposure of the Kaladgi group of sedimentary rocks located in the northern parts of the Karnataka state, are nearly 56 km long and 30 km in width (Fig. 1.11). The shape of the outcrop is approximately rectangular one. The shape of the outcrops of the Gondwana group of rocks is long either in E-W direction

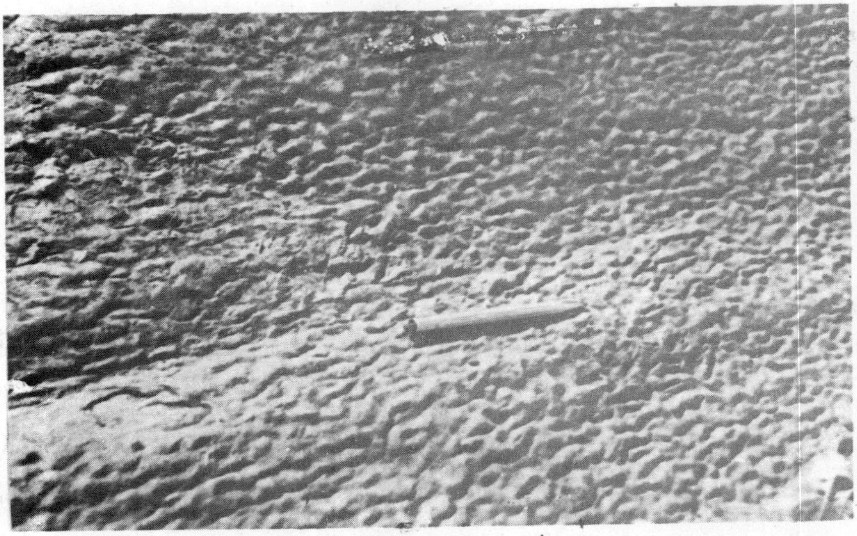

Fig. 1.10a. Excellently developed rain prints in the quartzarenites exposed 800 metres S 85° W of Lakhmapur village, Badami taluka, Bijapur district, Karnataka state. Note the elongated nature of rain prints, the ball-pen indicating such a direction. *Courtesy :* Dr. G.V. Hegde.

Fig. 1.10b. Excellently developed, unusually large sized rain prints in the quartzarenites exposed 5 km S 65° E of Shindogi village, Belgaum district, Karnataka state. Note the elongated shape of the cavities. Also note one complete almond being accommodated in one of the cavities (more central part of the photo). *Courtesy :* Dr. G.S. Pujar.

(Damodar, Sone and Narmada valleys) or in NW-SE direction (Godavari valley) (Fig. 1.12). Krishnan (1968) writes "These long and narrow tracts are really a series of faulted troughs" (1968, p. 240). This feature is once again suggestive of a rectangular shape of outcrop for the Gondwana rocks. Tethys which is by far the largest sedimentary basin, it once extended from the Atlas mountains and the Pyrenees in the west, through Alps, Carpathians, Caucasus, Himalayas and farther east into Burma-Arakan-Andamans. This basin certainly is very long than wide in its shape, the general trend being E-W. Outcrops of some other sedimentary rocks are oval or crescent-shaped, like the Cuddapah group of sedimentary rocks (Fig. 1.13). Ancient sedimentary basins also display a regular shape. Thus the Dharwar-Shimoga schist belt, the Gadag-Shrirangapattan schist belt, the Sandur schist belt, all of them decidedly display a rectangular shape, being more long in the NW-SE direction than across it (Fig. 1.14). These schist belts probably were once long and narrow basins in which sediments were laid. Therefore it may be said that the outcrop pattern and their shapes are certainly distinctive ones as compared to that displayed by the igneous or the metamorphic rocks. The shapes of the outcrops and the extent of the sedimentary rocks are solely controlled by that of the basin in which the sediments were laid.

The aspect to be emphasized from the disposition (shape of the outcrops) of the sedimentary rocks in the crust of the earth is that these constitute separate and individual units (like the Kaladgi, the Vindhyan etc.). Each unit inherits the dimensions (length; breadth and depth) of the basin. It is therefore possible to ascertain the exact extent of the unit with a fair degree of accuracy. This feature is especially useful in calculating the "coal reserves" (from coal seams), that of lime-

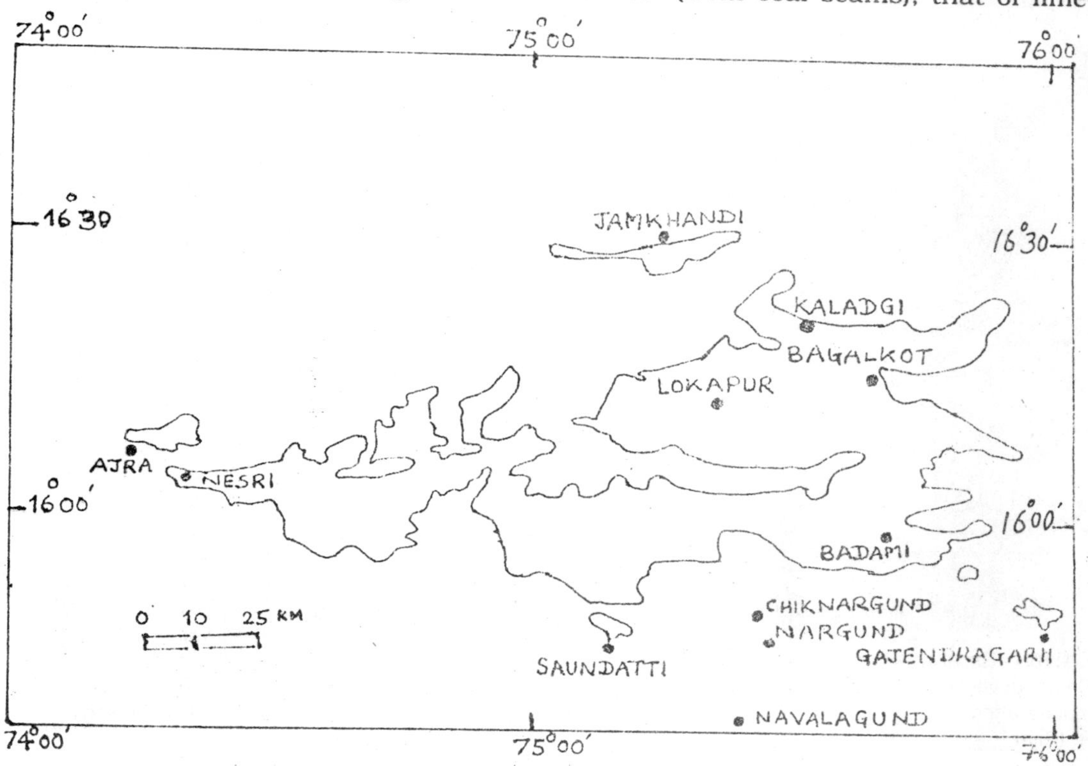

Fig. 1.11. Outcrop of Kaladgi (basin) group of rocks.

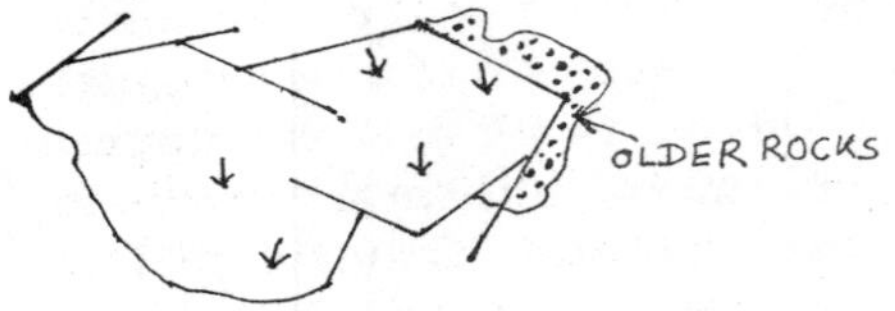

Fig. 1.12. Gondwana outcrop (basin) (a sketch map). Dotted area is of older rocks. Arrows indicate southerly dip of gondwana rocks (after Fig. 16 of Wadia, D.N. Geology of India, 1981, 4th edition).

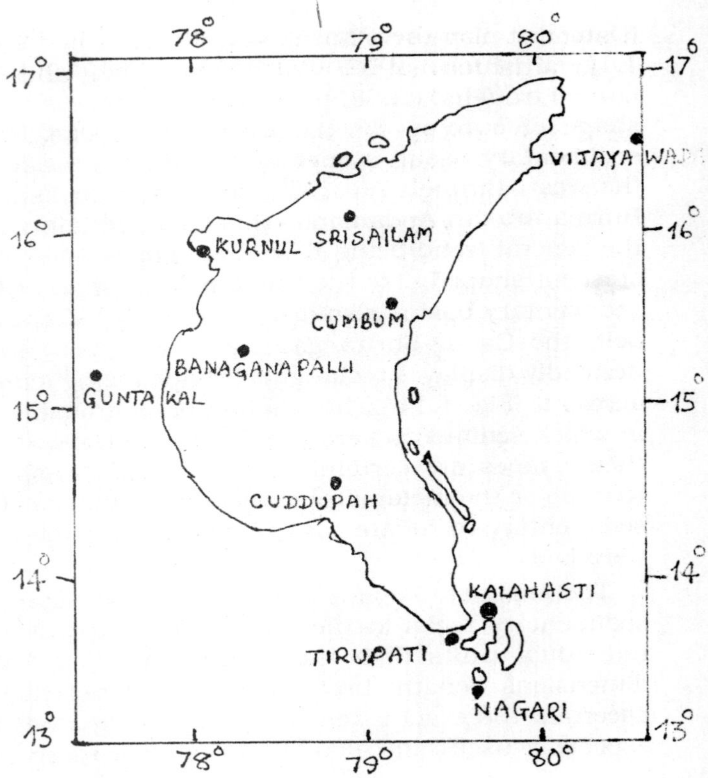

Fig. 1.13. Crescent shaped outcrop of cuddapah (basin) rocks (after map VIII of Krishnan, M.S. Geology of India and Burma, 5th edn., 1968).

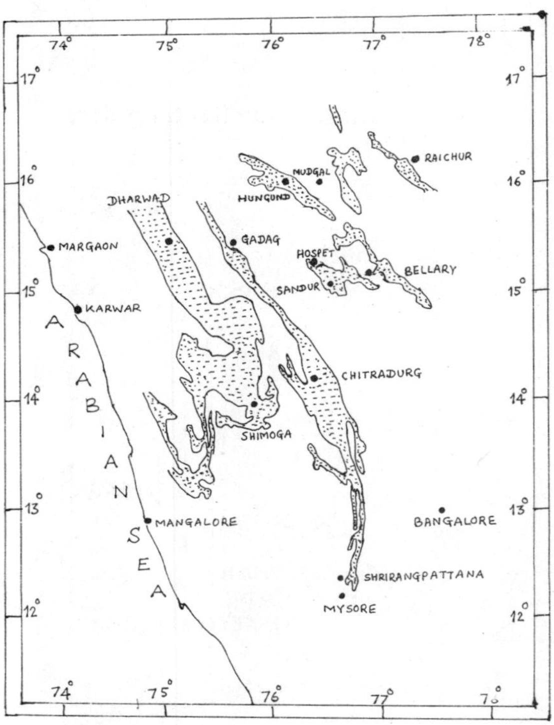

Fig. 1.14. Outcrop patterns of ancient schistose groups (basins) : Dharwar-Shimoga belt, Gadag-Chitradurga belt, Sandur schist belt, Hungund schist belt (after map VI of Krishnan, M.S. Geology of India and Burma, 4th edn. 1968).

stones etc. It may therefore be said that the mode of occurrence of the sedimentary rocks is in the form of "tabular bodies". Shape of the outcrop is the manifestation of it.

As is well known, geology is mostly a field science. Prior to setting out for field mapping, the geologist desires to get some idea about the kind of the rocks (whether igneous, metamorphic or sedimentary) that are likely to be encountered in the area selected for study. The topographical maps convey the nature of the hills and the valleys. As has been described earlier, the geomorphic forms derived from the sedimentary rocks are quite distinctive ones (for greater details, refer Dake and Brown, 1925). These days aerial photographs and the Landsat imageries yield information in respect of the rocks of the crust of the earth. Distinction between the igneous, the sedimentary and the metamorphic rocks is made from such imageries, and the aerial photographs. Further the extent of the rocks (shape) is also inferred. This information thus indicates the likelihood of the presence of the sedimentary rocks in the area under investigation. In the actual field work, more details like the bedded nature, typical structural features like current bedding (Figs. 1.6a, b), graded bedding (Fig. 1.7), ripple marks (Figs. 1.8a, b, c), confirm the presence of these rocks. The mineral composition and the presence of cementing material for the grains, are to be established only in the laboratory studies through microscopic examination of thin sections of rock samples. The distinctive features elaborated in this chapter are of immense importance to a field geologist and a researcher, to make the first approximation in respect of the presence or otherwise of the sedimentary rocks in the area selected for the study.

SUMMARY

Below noted features are noteworthy :

1. Typical flat-topped hills or dip-slope escarpment scenery.
2. Mode of occurrence as tabular bodies of somewhat regular shape (rectangular, crescent etc.) and of estimatable dimensions.
3. Bedded, layered, banded, laminated nature.
4. Rounded to subrounded grains set in a matrix.
5. Presence of current bedding, graded bedding, ripple marks etc.
6. Presence of fossils (this is the monopoly of the sedimentary rocks.
7. Monomineralism.

CHAPTER 2

Textures and Structures of Sedimentary Rocks

INTRODUCTION

While describing the distinctive features of the sedimentary rocks, some of the terms used like current bedding, the graded bedding etc., have already been noted. It is desired in this chapter to deal with their classification, origin and their utility in the geological studies.

CLASSIFICATION

The texture of any rock is influenced by the shapes and the sizes of constituent grains (minerals). The disposition of the grains in a rock produces a variety of appearances. The term texture is used when the inter-relations between the constituent grains are taken into account. But when the grains occurring in a group, and their disposition is considered, then the term structure is applied. A few examples will clarify the difference between a texture and a structure.

It is customary to consider the actual sizes of the grains based on some conventions agreed upon by the sedimentologists. Accordingly, fine, medium and coarse grain sizes are recognised. Categorisation into fine, medium or coarse grained varieties of rocks is considered as the texture of the rocks, because here the relation of the individual grains with each other, is used as a basis. It may so happen that in a rock exposure some layers might be fine grained, the adjacent layer might be coarse grained, while the layer next to it, might be fine grained in nature. In such a situation, the overall grain size of the group of grains is considered, and not of the individual grains. Therefore such appearances are described by the term structure. As an outcome of such a disposition, the rocks give rise to a "layered or a bedded" structure, and not to a texture. Grain size might be same, but the colour of the grains composing the different layers or beds, might vary. Such a situation then gives rise to a "banded" structure and not to a texture, because, the beds or the layers are produced due to a variation of colour of the constituent grains of the different layers or beds. The feature to be noted is that the inter-relation between the grains composing the different beds, is not taken into account, and therefore, the sedimentary rocks mostly possess numerous structural and a few textural varieties.

BASES OF CLASSIFICATION

Four bases are used while classifying the structures and the textures encountered in the sedimentary rocks. The bases are :

1. The grain size.
2. The shape of the grains.

3. The colour of the grains.
4. The depth of water in which the sediments (grains) were deposited.

Each basis is further divisible. This has been presented in Table 2.1.

Table 2.1. Bases of classification of textures/structures of sedimentary rocks

	Grain size		Shape of grains	Colour of grains			Depth of water		Chemical activity and action of organisms
	Disposition of grains	Actual grain size		Bands	Spots	Veins	Shallow currents generated	Exposure to atmosphere	
	Bedded structures								
Name of texture/structure	Laminated, current, false, cross, torrential, graded beddings	Coarse, medium, fine grained	Rudaceous, gritty, structures	Bands or beds	Spotted structure	Dendritic markings ramifying veins	Current, oscillation ripple marks	Rain, prints, mud/sun cracks	Trails, tracks, vugs, oolites etc.

CLASSIFICATION OF GRAIN SIZES

The grains of the sedimentary rocks not only vary considerably in size, but are also very irregular in shape. Though it is customary to describe the grains as rounded or subrounded ones, these in fact are not rounded in the geometrical sense. A grain identifiable without the aid of a pocket lens, is considered as coarse. But then the constituents of a conglomerate are so large and big in size that it does not sound proper to merely describe them as very, very coarse and as very big. A fine-grained sandstone will be observed to be certainly coarser than a shale. Therefore in order to classify the different sizes properly, limiting values for each category have been agreed upon by the sedimentologists. The different categories of grain sizes are named according to the shape of the grains. The classification of the grain sizes is presented in Table 2.2.

Table 2.2. Grain size categories and their names

Size (in mm)	Shapes of grains	
	Rounded/subrounded/sub-angular	Angular
> 256	Boulder	Block
64 to 256	Cobble	Rubble
4 to 64	Pebble	Breccia
2 to 4	Granule	—
1/16 to 2	Sand	Grit
1/16 to 1/256	Silt	—
< 1/256	Clay	—

Data from Pettijohn 1984, Ind. Edition, p. 30.

It will be observed from Table 2.2 that all the grains having sizes between 2 and 1/16 mm are called as sand; grains of silt range between 1/16 and 1/256 mm and so on. Each category, therefore, could be further split into fine, medium and coarse sub-divisions. The process of sedimentation is a continuous phenomenon. Therefore the change of grain size cannot be expected to be sharp and abrupt, but to be a gradual one. Such a behaviour will affect the grain size of the several beds laid in a basin. The sedimentologists therefore have introduced a "grade scale" to measure the size of the grains composing the sedimentary beds. Many scales are in vogue like that of Udden (1898), Attenberg (1905), Krumbein (1934) and so on. The phi scale of Krumbein (op. cit.) is widely used. phi is the exponent (logarithm to the base 2) of the diameter of the grains. In order to avoid negative numbers, the log is multiplied by minus 1. The phi scale is equal to "minus $\log_2$ diameter in mm". The phi scale of Krumbein (op. cit.) and the class limits of Wentworth-Lane (Wentworth, 1922, Lane et al. 1947) which are widely used, are presented in Table 2.3.

Table 2.3. phi values, class limits and general terms (after Pettijohn, 1984, Fig. 3.6, p. 35)

phi values (in mm)	Class limits of Wentworth-Lane (in mm)	General terms
Minus 8 to minus 9	256 to 512 and above	Boulder
Minus 7 to minus 8	128 to 256	Large cobble
Minus 6 to minus 7	64 to 128	Small cobble
Minus 5 to minus 6	32 to 64	Very coarse gravel
Minus 4 to minus 5	16 to 32	Coarse gravel
Minus 3 to minus 4	8 to 16	Medium gravel
Minus 2 to minus 3	4 to 8	Fine gravel
Minus 1 to minus 2	2 to 4	Very fine gravel
0 to minus 1	1 to 2	Very coarse sand
0 to plus 1	1 to 1/2	Coarse sand
Plus 1 to plus 2	1/2 to 1/4	Medium sand
Plus 2 to plus 3	1/4 to 1/8	Fine sand
Plus 3 to plus 4	1/8 to 1/16	Very fine sand
Plus 4 to plus 5	1/16 to 1/32	Coarse silt
Plus 5 to plus 6	1/32 to 1/64	Medium silt
Plus 6 to plus 7	1/64 to 1/128	Fine silt
Plus 7 to plus 8	1/128 to 1/256	Very fine silt
Plus 8 to plus 9	1/256 to 1/512	Coarse clay
Plus 9 to plus 10	1/512 to 1/1024	Medium clay
More than 10	More than 1/1024	Fine clay

The gravels are also called as pebbles

Study of Table 2.3 indicates that there is a considerable range between the class-limits set out for the "sand, the silt and the clay" categories. A grains size varying between 2 and 1/16 mm is generally classified as sand. Therefore it is logical to describe a grain of 1/16 to be "very fine". Hence in each category, "sand, silt or clay", sub-divisions of "coarse, medium and fine" are made. These general terms are more popular and are in vogue, because these are readily adoptable during the field studies. The values for each sub-division are shown below.

Coarse — 1 mm and above. Grain (minerals) are identifiable without the aid of a pocket lens.

Medium — Less than 1 mm. Grains (minerals) identifiable with a pocket lens.

Fine — Grains (minerals) identifiable only in thin section of rock under a petrological microscope.

CLASSIFICATION OF SHAPES OF GRAINS

Unlike size, the shapes of the grains of the sedimentary rocks do not display much of variation. Owing to the wear and tear of the sediments (grains) brought about during the process of transportation, the angularities or the edges of the grains get smoothened and rounded, and the grains tend to become ellipsoidal or spherical (rounded) in shape. The shape also depends upon the inherent form (habit) of the minerals composing the rocks undergoing weathering, and on the type of the agent of transportation. Angular grains are either produced due to fracturing and rupturing of the parental rocks in situ, or due to a very short distance of transportation of the sediments by the geological agents.

Shapes of grains commonly encountered in the sedimentary rocks are rounded to subrounded, discoidal, spherical, bladed, rod-shaped, angular, sub-angular and so on. Roundness of a grain is calculated by the equation, and the same is given below.

$$\rho = \sum \frac{r_1/R}{N}$$

where r_1 = individual radii of the corners of the grains,
R = radius of the maximum inscribed circle, and
N = number of corners of the grain.
(See Fig. 2.1 for r_1 and R.)

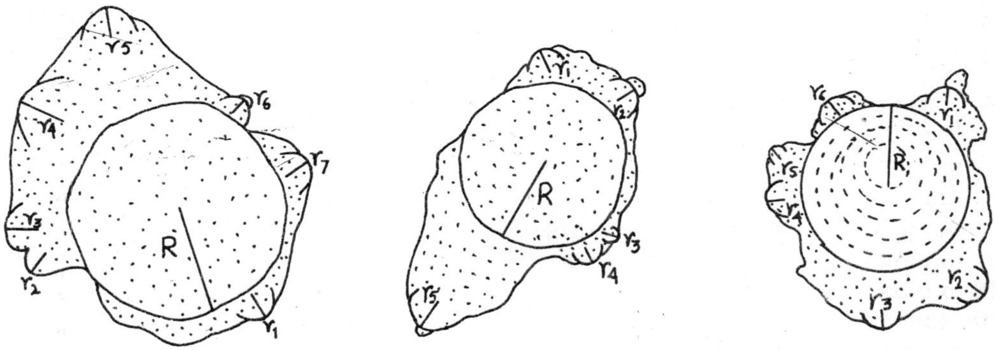

Fig. 2.1. Drawing of inscribed (r_1, r_2, r_3, ...) and circumscribed circles (R) within a grain (dotted area).

The word "roundness" is not to be mistaken for a spherical shape, but it means that the edges and the sharp corners of the grains are rounded off, and a smooth surface is developed. This has been shown in Figs. 2.2a, b, c, d, e, f. Based on the values of roundness, different names are given to the shapes displayed by the grains.

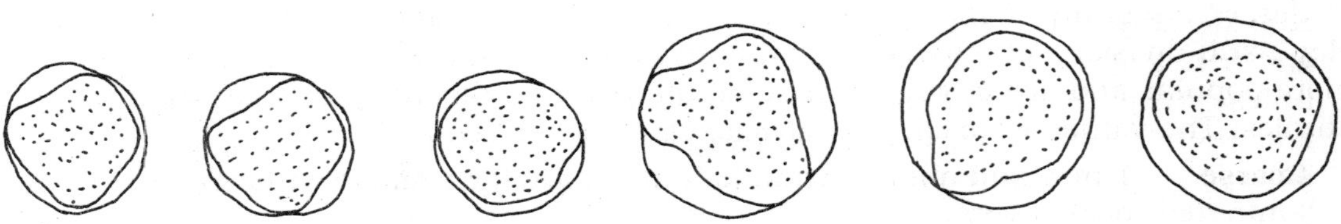

Fig. 2.2. Various shapes of roundness of grains (dotted area). Grains are about same size, but these differ in sphericity (grains taken from Ph.D. thesis of K. Bhimsen, 1989).

Pettijohn's values for grades of roundness and the names of the shapes are set out in Table 2.4.

Table 2.4. Roundness values and names of grade (shapes)

Value (ρ)	Grade term (shape)
0.00 to 0.15	Angular
0.15 to 0.25	Sub-angular
0.25 to 0.40	Sub-rounded
0.40 to 0.60	Rounded
0.60 to 1.00	Well rounded

CLASSIFICATION OF STRUCTURES BASED ON DISPOSITION OF GRAINS, GRAIN SIZES AND THEIR COLOUR

This basis practically covers almost all of the structures possessed by the sedimentary rocks. The actual size and the distribution of such grains gives rise to the different varieties. Various structures thus produced are set out in Table 2.5.

Table 2.5. Structures produced from size and colour of grains

	Disposition of grains	Colour of grains		Actual grain size
Name of the texture/structure	Bedded structures, laminated structures, current/false/cross-bedded structures, torrential bedded-structures, graded bedded-structures	Development of coloured beds/bands	Spotted structure	Coarse grained Medium grained Fine grained

Generally the sandstones display a coarse grained structure, while the shales and the limestones are normally fine grained rocks. Participation of the coarse, the medium and the fine grains gives rise to a variety of structures. Development of the bedding planes is the first outcome of it. Across such planes, the grain size is different, and therefore two different bodies (later called as beds) are produced. The grain size may vary several times during the process of sedimentation. As a result of this, sediments varying in grain size are laid one over another in a conformable manner, which after compaction, gives rise to several beds (Fig. 1.1a, b, c and the frontispiece photo).

Sandstones are the most extensively formed rocks throughout the geological history of the earth. These rocks characteristically display bedded nature more often than the other (shales, limestones) sedimentary rocks. The development of beds is an unique and a predominant structure of these rocks. When the beds are considerably thin and of alternating very fine grain size, and are laid one over another, it gives rise to the "laminated structure" (Figs. 2.3a, b). It is argued that unless the grain size of each layer is fine, the individual layers cannot become thin, and therefore in turn, it cannot give rise to very thin laminations. The structure however is due to the pilling up of a large number of very thin layers. Such thinly laminated rocks when hammered, are prone to cleave along the bedding planes. This behaviour called as "bedding fissility" therefore simulates (resembles) the rock-cleavage, the latter one being due to the action of the deformative forces on the pre-existing rocks, and is not the outcome of sedimentation.

Generally the several beds (layers) are laid one over another in almost horizontal planes. However at times these are deposited at an angle to the beds already laid at the bottom. This gives rise to the "current bedded structure" (Figs. 1.6a, b and 2.4). This structure is technically split into the top-set, the off-set and the bottom-set beds

Fig. 2.3a. Sandstones exposed in a quarry near Chikkashellikeri, Bijapur district Karnataka state. Note the thinly laminated nature of the sandstones, unlike the thick beds seen in Figs. 1a and b. *Courtesy* : Dr. M.J. Chandrasekhara Gowda.

Fig. 2.3b. Quartzarenites exposed near Badami town, Bijapur district, Karnataka state. Note the laminated nature of the rock as well as the colour banding. *Courtesy* : Dr. G.S. Pujar.

(Fig. 2.4). Important feature to be noted is the angle of repose of the off-set beds with the bottom-set and the top-set beds. Towards the bottom-set beds, the angle is gentle and has a somewhat asymptotic relationship. At the top-set beds, it is quite high and abrupt (Figs. 1.6b and 2.4). Current bedding is also known as "false/cross/bedding". These are synonymous terms. Whereas the top-set and the bottom-set beds are horizontal, those in between them are inclined. These are therefore called as the off-set beds. When the off-set beds are highly inclined or are nearly perpendicular to the top-set and the bottom-set beds, this disposition gives rise to the "torrential bedding" (Figs. 2.5 and 2.6). At times the grain size varies from coarse at the bottom, to fine towards the top in a gradual manner. Such a structure is described as the "graded bedding" (Figs. 1.7 and 2.7).

Depending upon the actual grain size, "coarse, medium and fine grained rocks" are formed. As has already been noted, the sandstones are generally coarse to

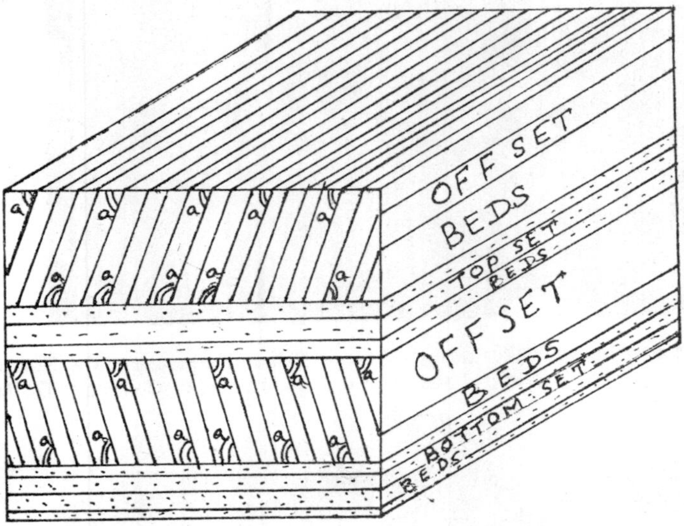

Fig. 2.4. Relation between offset, topset and bottomset beds of a current bedded structure. Note that the topset and bottomset beds are horizontal, while the offset beds are inclined. Offset beds display "acute truncation" (angle "a" in the figure) towards the "top set beds", while these display a "gentle surface of contact" (angle "b" in the figure) at the "bottom set beds". Also note that the "topset beds" of the "lower current bedded" structure, assume the status of "bottom set beds" for the "upper current bedded" structure. Current bedding is also called as "cross/false" bedding. The top- and the bottomset beds are shown with dots.

Fig. 2.5. Hand-specimen of a quartzarenite showing torrential bedding in the lower part (palm of the person), and the current bedding on the upper part. Note the sharp line of contact between the two structures. Rock specimen is from Khanapur, Badami taluka, Bijapur district, Karnataka state. *Courtesy :* Dr. G.V. Hegde.

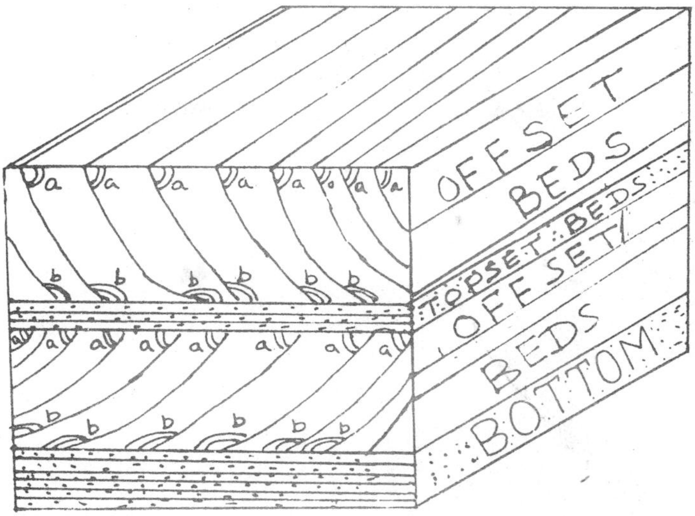

Fig. 2.6. Relation between topset, offset and bottomset beds in a torrential bedded structure. Note that the top- and the bottomset beds are horizontal, while the offset beds display high angle of inclination (angle "a" in the figure) towards both "topset and bottomset" beds. Also note that the "topset beds" of the "lower torrential bedded" structure assume the status of the "bottomset" beds for the "upper torrential bedded" structure. The top and bottomset beds are shown with dots.

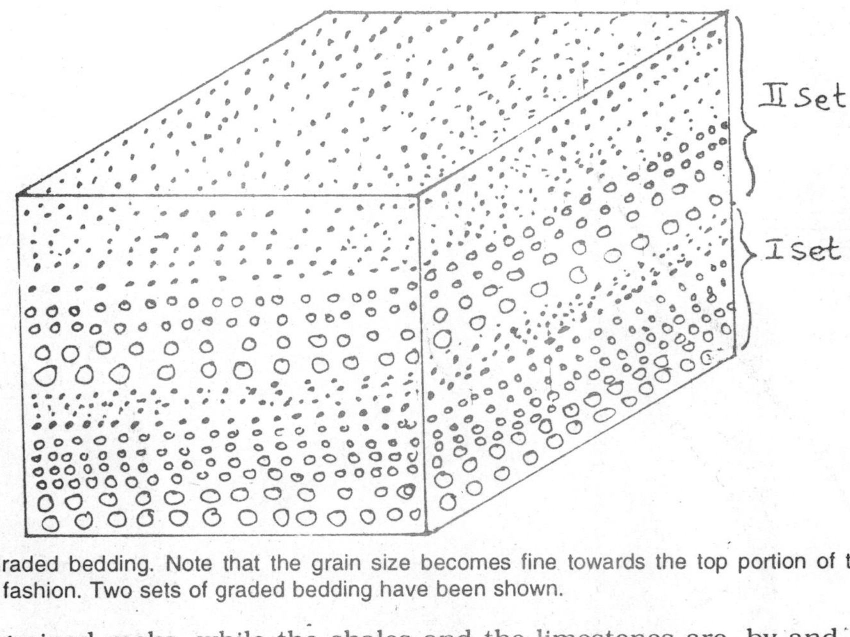

Fig. 2.7. Graded bedding. Note that the grain size becomes fine towards the top portion of the rock, in a gradational fashion. Two sets of graded bedding have been shown.

medium grained rocks, while the shales and the limestones are, by and large, fine grained in texture. Sometimes the dimensions of the grains are so large that it cannot be merely called as a grain at all. In such instances, the term pebble, cobble or boulder etc., are then made use of. It is to be noted that though the sizes are enormous, the individual pebbles, cobbles or the boulders, are rounded to sub-rounded in shape. These are also held in a finer grained matrix. This set-up gives rise to the "rudaceous structure" (Fig. 2.8a, b), and it is displayed by a conglomerate.

Thus it may be said that based on the grain size, structures recognised are bedding planes, current/false/cross/torrential bedding, graded bedding, rudaceous structure, coarse/medium/fine grained textures and so on.

Fig. 2.8a. Excellently exposed thick conglomerate bed forming a part of the Bannikoppa-Bagewadi horizon from the Gadag schist belt, Dharwad district, Karnataka state. Note the rounded to sub-rounded shape and smooth surface of the pebbles, and the large sizes of the boulders and the pebbles. The bed is dipping 60-70° due southeast. The dimensions of the pebbles and the boulders are so much large that the rock constitutes a "boulder bed". The rock displays characteristic rudaceous structure. *Courtesy :* Dr. N.N. Gothe.

Fig. 2.8b. Rudaceous structure displayed by the conglomerate horizon exposed near Kardikolla (4 km S 40° W of Vadrahalli village, Bellary district, Karnataka state). Note the large but uniform sizes of the pebbles. The rock possesses a high angle southerly dip. *Courtesy :* Dr. H.D. Desai.

At times the bedded structure is formed through a variation in the colour of the grains composing the rocks. Depending upon the colour of the grains, beds of pink, violet, gray, white or cream coloured rocks are formed. Sandstones exposed in the vicinity of Gorvanakolla village, Belgaum district, Karnataka, are pink in colour (Fig. 2.9 and the frontispiece photo). However there are horizons in between them that are violet in colour (Fig. 2.10). Even gray coloured varieties are also met with, in the said area. The Red Fort of Delhi is so named because of the red coloured sandstones used in its construction. Sometimes the rocks are not coloured uniformly. Hegde (1984) and Pujar (1989) have reported occurrences of coloured spots in the sandstones from the southeastern parts of the Kaladgi sedimentary basin, Karnataka state. The spots are due to the presence of violet or gray coloured aggregate grains of quartz. The pink or the violet coloured sandstones are observed to contain gray spots, while the gray coloured sandstones contain violet/pink coloured spots. Hegde (1984) and Pujar (1989) have reported (colour) spotted sandstones from the southeastern parts of the Kaladgi sedimentary rocks. There are violet spots in gray coloured sandstone (Fig. 2.11). At places some spots display a diffused border (Fig. 2.12), and these have imparted an uniform violitish tinge to the entire rock. Hegde (op. cit.) reports sandstones displaying violitish coloured streaks penetrating the inner part of the rock (Fig. 2.13a). At about 2 km east of Jamkhandi town, Bijapur district Karnataka state, sandstones possessing veins and veinlets of violitish coloured material are exposed (Fig. 2.13b). At times, coloured grains get restricted to the

Fig. 2.9. Pink quartzarenite exposed at about 2 km northeast of Gorvankolla village, Belgaum district, Karnataka state. Also note a large sized purple coloured spot that shows diffused borders. To the immediate right hand side of the spot, another faint spot is noticeable, the feeble colour being due to absorption of the spot by the grains surrounding it. Also noticeable is "Percussion structure (radial joints)". *Courtesy :* Dr. G.S. Pujar.

Fig. 2.10. Exposure of a violitish coloured quartzarenite in the eastern vicinity of the Malaprabha dam site (the famous peacock gorge), Saundatti taluka, Belgaum district, Karnataka state. Also note the occurrence of pink quartzarenite layer towards the bottom part of the outcrop. Also note the occurrence of 2 horizons of intercalated conglomerate, each layer being 4-6 cm thick. *Courtesy* : Dr. G.S. Pujar.

Fig. 2.11 Hand-specimen displaying "spotted structure", developed in the quartzarenites of Belavalkop area, Badami taluka, Bijapur district, Karnataka state. The spots are oval to nearly circular in cross-section, light pink in colour, and are disseminated throughout the rock. The spots consist of violet coloured segrega[illegible] grains of quartz. *Courtesy* : Dr. G.V. Hegde.

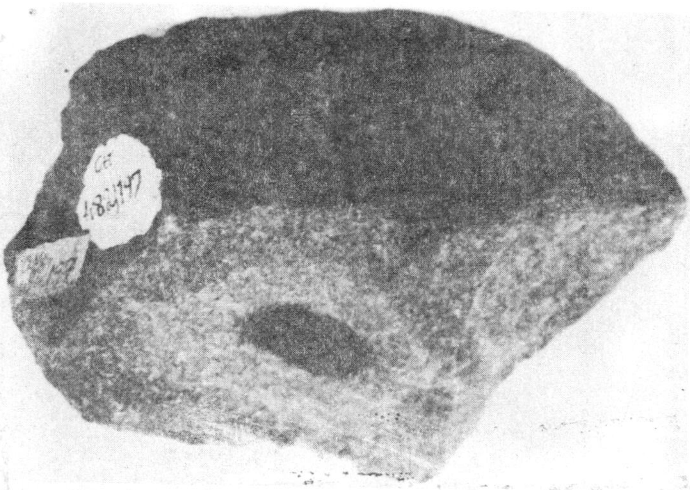

Fig. 2.12. Hand-specimen of a pinkish brown quartzarenite containing ovoidal shaped ferruginous inclusion (the darkest tone). It is 2.5 × 1.5 cm in dimension, and is surrounded by two coloured zones. The first is honey coloured, while the outermost is purple in colour. The specimen is from Mudakavi area, Belgaum district, Karnataka state. *Courtesy* : Dr. G.V. Hegde.

Fig. 2.13a. Light gray coloured sandstones showing abrupt termination of bedding planes against a non-bedded sandstone of different colour (block of rock on the right hand top side of the photo). The dark coloured thin band in the blocks of rock is a band of pink grains of quartz. Note that the dark band is not situated on the surface of the rock, but it further takes a 90° bend (see the large blocks of rock on the left hand side of the photo) in its course. This structure is encountered in the northern vicinity of the Ugalvat village, Bijapur district, Karnataka state. *Courtesy* : Dr. G.V. Hegde.

bedding plane of the rock, and give rise to dendritic markings (Fig. 2.13c). All these instances may be categorised as "unusual structures" produced by the coloured grains of quartz.

Fig. 2.13b. Field photo of pink quartzarenites traversed by darker coloured veins, in a criss-cross manner. The veins consist of purple coloured aggregated grains of quartz. Locality is Jamkhandi town, Bijapur district, Karnataka state. Photo taken by the author.

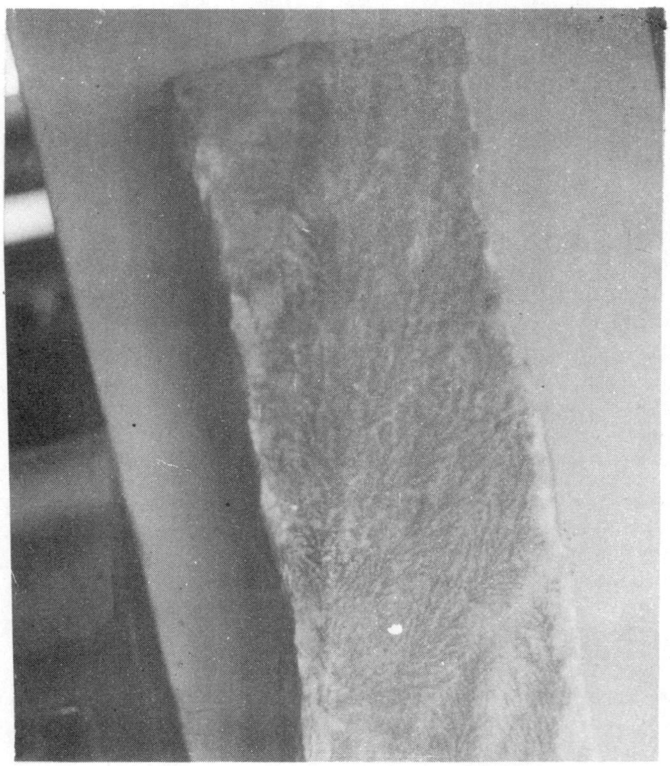

Fig. 2.13c. A large slab of light cream coloured sandstone displaying excellently developed dendritic markings. *Courtesy :* Department of Studies in Geology, Karnatak University, Dharwad.

CLASSIFICATION OF TEXTURES BASED ON SHAPES OF GRAINS

As compared to the grain size, the shapes of grains give rise to a few textures. When the grains are big and enormous, and are neither rounded, subrounded, nor angular in shape, it produces the "gritty texture", the host rock itself being called as a grit (Fig. 2.14). When the grains are angular or irregular in shape, a "breccia structure" is developed (Figs. 2.15a, b, c). However this structure should not be mistaken for a "fault breccia, or a volcanic breccia" because there is a considerable resemblance between them.

CLASSIFICATION OF STRUCTURES BASED ON DEPTH OF WATER IN THE BASIN

When the depth of water is shallow, the waves and the currents can reach the sediments laid in the basin. It is also possible that the sediments might get exposed to the atmosphere, especially when the depth of water is quite shallow. Structures produced under such an environment are the ripple marks, the rain prints and so on. These structures are readily identifiable in the field studies. Ripple marks are further divided into oscillation and current ripples (Figs. 1.8a, b, c and 2.16a, b). These are fantasies in the sedimentary rocks. The shape of the wave produced in the water is imprinted faithfully onto the sediments laid in the basin. The waves are caused by the onward movement (current ripples), and by the up and down movement of water (oscillation ripples). The current and the oscillation ripple marks are encountered more commonly in the medium and the fine grained sandstones. Size of a ripple is measured by the height and the length of the wave. When these two

Fig. 2.14. Field photo of arenaceous rock showing gritty texture. Note the presence of uniform sized and shaped numerous grains of quartz disseminated throughout the specimen. Outcrop of this rock is encountered on the Belgaum-Nesargi road. *Courtesy :* Dr. S.C. Puranik.

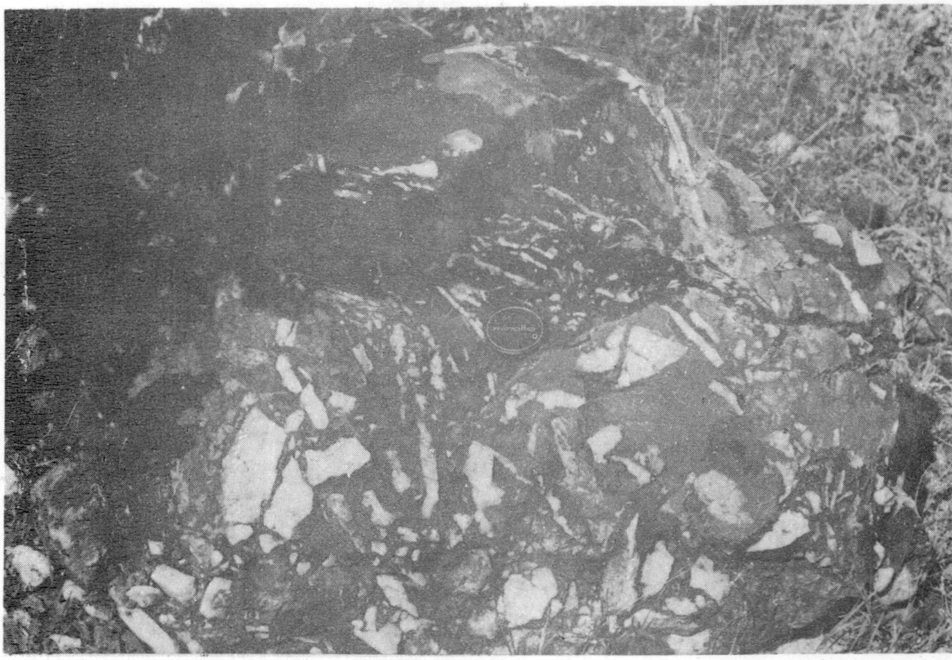

Fig. 2.15a. Field photo of arenaceous rock showing breccia structure. Note the angularity of the pieces and their irregular sizes. The structure is noticeable near Aralikatti village, Lokapur, Bijapur district, Karnataka state. *Courtesy :* Dr. M.J. Chandrasekhara Gowda.

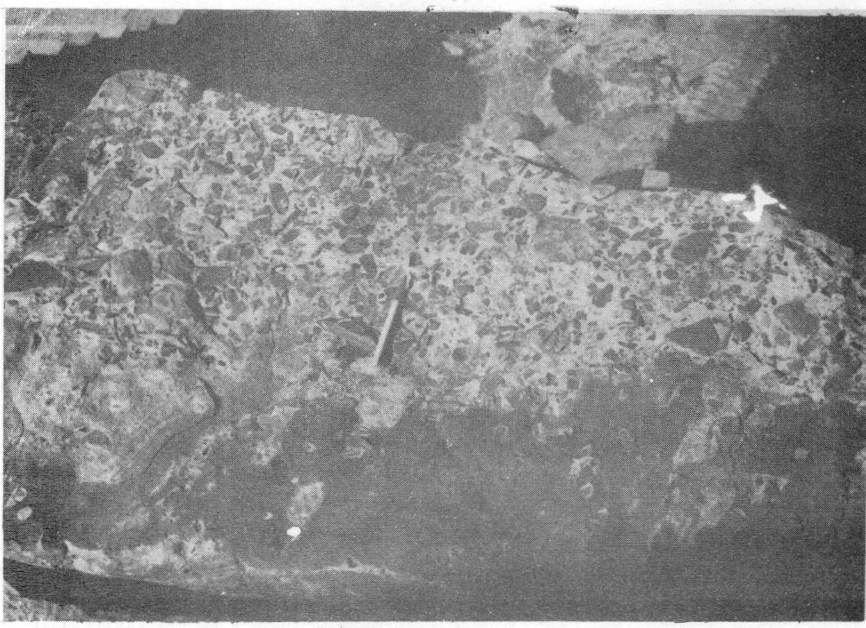

Fig. 2.15b. Field photo of excellently developed breccia in the vicinity of Aihole, Bijapur district, Karnataka state. *Courtesy :* Dr. S.C. Puranik.

Fig. 2.15c. Field photo showing the development of a breccia-cum-conglomerate in the southwestern vicinity of Kamewadi village, Daddi, Belgaum district, Karnataka state. Note a great variation in the sizes, and the irregular shapes of blocks and pieces of rock. *Courtesy :* Dr. K. Bhimsen.

dimensions are remarkably large, it gives rise to "giant ripples". Bhimsen (1989) has reported one such instance which is presented in Fig. 1.8c.

In the shallow water basins, there is the possibility of the sediments getting exposed to the atmosphere. As a result of such an exposure, the wet sediments start drying up and cracks are produced in them. The fantastic aspect is that such cracks are faithfully preserved in tact, even after the compaction of the sediments into hard sedimentary rocks. These structures are called as the "mud or sun cracks". This structure is common in the clay rocks. During the period of exposure to the atmosphere, the area might receive showers of rains. These might create tiny holes and depressions in the soft sediments. Fantastic part is that such holes are often times perfectly retained by the rocks formed from such sediments. The tiny holes or depressions give rise to the structure called as the "rain prints" (Figs. 1.10a, b).

ORIGIN OF TEXTURES AND STRUCTURES

The textures and the structures reflect the process of formation of the concerned rock, whether igneous, sedimentary or metamorphic. The grains called as the sediments in the sedimentary rocks, hold a clue to the understanding of the process or the processes by which these rocks had originated. The size of a grain is correlated to the process of crystallisation or of recrystallisation in the cases of the igneous and the metamorphic rocks, respectively. However in the case of the sedimentary rocks, it is dependent on the source rocks, and on the amount of transportation undergone by the sediments. There is one more factor to be considered. The sediments are brought into the basin, and these later on settle down to the bottom of the basin. The depth of water in the basin is therefore yet another factor to be taken into

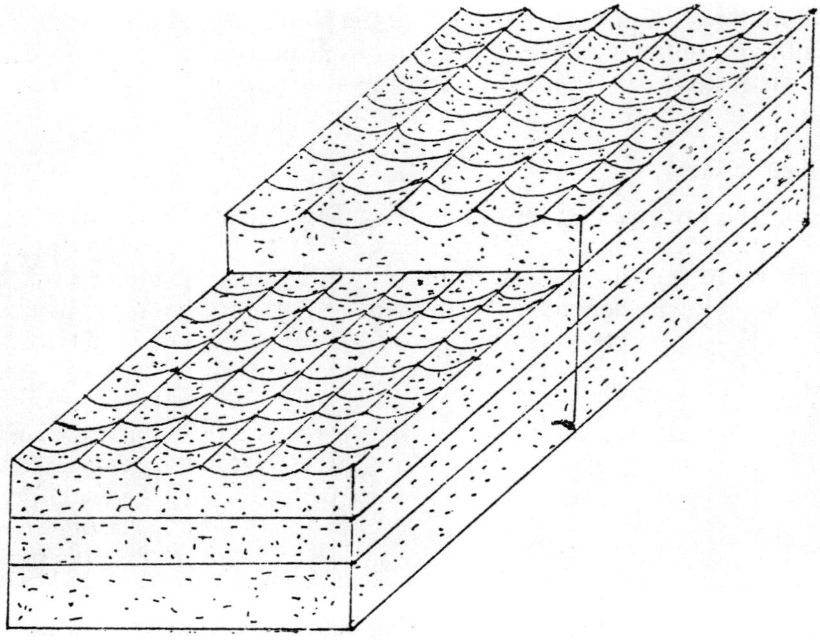

Fig. 2.16a. Oscillation ripple marks. Note that the concavity of ripples is upwards or towards the topside of the bed. Two ripple marked surfaces are shown. Also note that the ripples are symmetrical in shape. This structure is produced only in shallow depth of water through up and down movement of water. This structure is also produced only in shallow water.

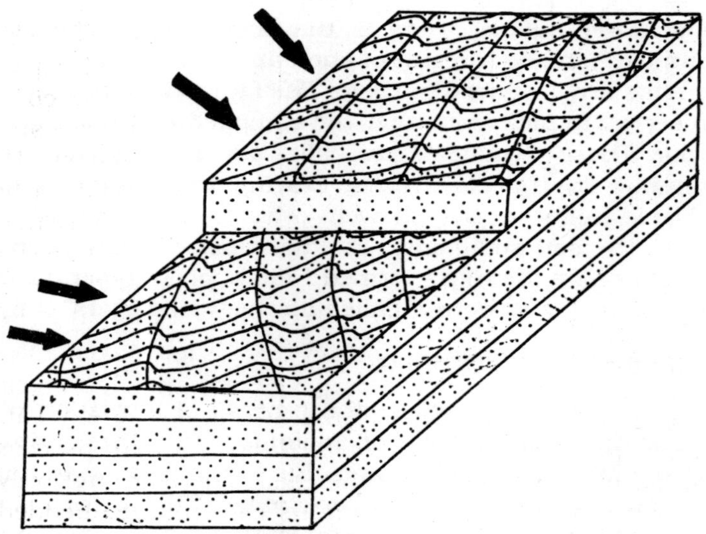

Fig. 2.16b. Current ripple marks. Note that the wave form is asymmetrical in shape (compare Fig. 2.16a). It is more long in the direction of flow of the current. Two ripple marked surfaces have been shown. Disposition of the current ripples is used to obtain the direction in which the sediments were brought into the basin. Curved lines (these are numerous) mark the wave surfaces. The other rather straighter lines (these are few), mark the axes of the wave surfaces. Arrows indicate the direction of flow of the currents. Note that the direction of currents is different for the two surfaces.

account. If shallow, certain structures like the ripple marks etc., are produced. When deep, possibly more layers or beds differing in grain size might be formed, giving rise to bedded structures. In the light of such a situation, the origin of the different textures and the structures are required to be understood.

(a) Bedded structures

These structures are predominantly caused due to a marked variation in the grain size of the sediments brought into the basin by the transporting agents. The grain size depends upon the distance of transport undergone by the sediments, as well as the nature of rocks undergoing weathering—whether hard, soft and so on. It has been held by the geologists that the majority of the sediments are transported by the streams and the rivers. Longer the travel, greater will be the wear and tear of the sediments (grains). Therefore such of the rivers that are situated at far away regions with respect to the basin, will produce finer sized grains as compared to those that are situated relatively closer to the basin. Reduction of the grain size also depends upon the nature of the source rocks (called as the provenance in the sedimentary terminology) undergoing weathering. Harder ones will produce coarser grains (sediments) than the softer varieties of rocks. Another point of importance is that the sedimentary rocks produced by an earlier geological episode of sedimentation, might be constituting a provenance for the succeeding episode of sedimentation. The grains of such rocks could be already finer ones, as well as rounded in shape. Thus while inferring the distance of transport undergone by the sediments, this aspect should be taken into account. Wind also is an active agent of transportation. Development of the finer or the coarser grains will depend upon how much strong were the winds blowing. It has been observed that the grains transported by the wind are very much fine and more rounded ones. These grains are called as the "millet seed sands".

The rivers and the wind bring the sediments to the site of deposition namely the basin. The rate of settling of the finer, the medium and the coarser grains being different, distinct layers or beds corresponding to the said three sizes, are produced. The coarser grains settle down first being relatively more heavy. This is followed by the medium and the finer categories of grains. This results in either three different layers, or a gradation of the grain size will be produced. As already noted, when the layers are very thin, and the grain size is very fine, it gives rise to the "laminated structure". It may also be designated as "thinly bedded structure" (Figs. 2.3a, b). In this manner the bedded structure is acquired by the sedimentary rocks.

As already noted, the sedimentary rocks vary in their colour like red, pink, white, cream colour, gray and so on. How they acquire different colours is a debatable issue. At present it is sufficient to note that if in a succession of say sandstones, colours of the grains vary from bottom to the top, then this may produce layers or beds differing in colour from one bed to another bed. Therefore not only the grain sizes, but even the colour composition of the grains can produce bedded structures (Fig. 2.17).

Each bed may be considered as a separate and a distinct period of settling down of the sediments of a particular grain size. Afterwards there could be a pause when sediments might not be settling down. This period of "non-sedimentation" might be followed by a fresh quantum of sediments settling onto the already deposited layer. These two layers might not necessarily differ from one another in respect of grain

Fig. 2.17. Field photo showing development of purple coloured banding in pink coloured quartzarenites. Palm of the person marks an area where a large purple spot is clearly noticeable. In the field it looks as though bands have been drawn in "blue ink" on the pink coloured quartzarenitic host rock. *Courtesy* : Dr. G.S. Pujar.

size or in colour, yet a surface of contact is created between them, which gives rise to a bedding plane.

The entire process of sedimentation is complex one. It is not clear as to whether all the sediments like coarse, fine, medium, and those carried in solution (like Ca, Mg) are brought into the basin, at one and the same time. If so, only then the sediments could be deposited (settled) one over another, progressively becoming finer towards the top. But the point is that this cannot be said with certainty. It is also possible that the rivers carrying (transporting) sediments, as they enter the basin, the coarsest sediments will get deposited (first) close to the margin of the basin. The finer ones might be getting carried farther into the deeper and the inner parts of the basin. Under such a situation, the layers (beds) cannot be expected to be laid one over another, because the sediments differing in grain size will be settling down at different depths of water, and therefore not at one place, or at the same place in the basin.

(b) Causes of variation in thicknesses of beds

The rivers are continuously at the work of depositing the sediments into the lakes, the seas and the oceans. During the rainy season, the load carried by them is naturally and definitely more than that carried during the other seasons. Therefore probably more quantity of sediments are laid during the flood times, giving rise to thicker beds. Those deposited during the other seasons could be expected to be less in quantity, and hence thinner layers or beds might be formed. As has been said earlier, more than one river might be bringing in sediments into the basin. Therefore it is likely that one river (initiated over softer rocks) brings more sediments, while another (if it be initiated over harder rocks) brings in less quantity of sediments. Such a situation will produce beds of unequal thicknesses in the basin.

Likewise the depth of water need not be necessarily same in all parts of the basin. The floor of the basin also could be uneven, thus producing shallower and deeper stretches within it. Such a situation also could give rise to varying thicknesses of beds; thicker in the deeper stretches, and thinner in the shallower ones. Even the grain size could vary; coarser in the shallower stretches, finer in the deeper stretches, and so on. These are all interpretations, because it is not possible to demonstrate the process of sedimentation in the laboratory.

(c) Current/false/cross/torrential bedding structures

These structures are originated in the sediments that are laid in the shallower depths of water. Current, false, and cross bedding structures are one and the same i.e., these are synonymous names. The off-set beds are found to be at an angle to the top- as well as to the bottom-set beds, these latter two beds being horizontal in disposition. It is held by the sedimentologists that the water in the basin was not quiet, and the inflowing river might have generated currents in the water. Such currents appear to have upset the sediments settling vertically down. Therefore, instead of producing horizontal layers, it gives rise to inclined ones. This period is followed by the normal, quiet condition of water favouring vertical settling down of the sediments, resulting in horizontally reposing beds. Study of current bedded structure reveals that these were repeated many times giving rise to several units of current bedded structures lying one over another in a vertical succession. There is one more feature that needs a mention. The direction of inclination of the off-set beds belonging to the different units, is not the same. It is towards left hand side for one set, while it is towards the right hand side for a set found on the overlying or the underlying off-set beds (Figs. 1.6a, and 2.3). This disposition clearly indicates that there is a continuous change in the direction of the currents developed in the water of the sedimentary basin. Though the exact modus operandi of the production of the current/false/cross bedding structure is a debatable matter, it may however be said that the currents generated in the water, and the continuation of the flow of the river into the basin, together are responsible for the development of the said structures.

In the case of torrential bedding (Figs. 2.5 and 2.6), the off-set beds are inclined at high angle or are almost perpendicular to the top-set and the bottom-set beds. The origin of this structure is once again attributed to the currents produced in the water of the basin. Probably the nature and/or the intensity of the currents generated in the water was different than those responsible for the development of the current bedded structure.

(d) Graded bedding

Characteristic feature of this structure is that the grain size becomes finer towards the top part of a bed, in a gradational manner (Figs. 1.7 and 2.7). Depending upon the sizes of the grains, the rate of settling of grains in water is different. Larger and bigger grains settle down faster, and therefore first, while the finer ones take more time and hence settle down slowly and much later. In between these two periods, grains of intermediate sizes settle down. Therefore the first grains to settle down will be coarse, while those to settle down last will be very fine. The grain size thus grades from coarse at the bottom, to very fine towards the top. To effect a separation of the different grain sizes, a quiet water, devoid of currents, is necessary. Such an environment is conducive to maintain differential rates of settling for the coarse, the medium and the fine grain sizes. Graded bedding is a case of "gravity settling" process, like the one operative in the case of crystallisation process of some gabbroic bodies (which have produced rhythmic layering structure in them).

(e) Spotted structure

This is an unusual structure and is displayed by some sandstone occurrences (Figs. 2.11 and 2.12). It is due to the presence of coloured spots, either gray or violet, in a relatively light coloured sandstone. The coloured spots are composed of aggregates of grains of quartz. As has been noted earlier, the pink or the violet coloured sandstones contain gray coloured spots, while the gray coloured sandstones contain violet coloured spots. The spotted structure appears to have a bearing upon the (colour) banded structure. While describing violet or the gray coloured sandstones, it has been noted that violet or coloured spots having diffused borders (Fig. 2.12) are encountered in some gray and violet coloured sandstones. How the grains of quartz have acquired gray or the violet colour is a debatable matter. It is likely that during the process of compaction of the sediments, some gray coloured beds were broken into big and small pieces, and these in turn got mixed up with the unbroken beds of sandstones. Likewise, some violet coloured beds were broken into big and small pieces, and these in turn got mixed up with the unbroken beds of sandstones. Logic behind tendering such a process rests on the fact that pink coloured sandstones are observed to contain violet coloured spots possessing diffused borders. Likewise, gray coloured sandstones contain gray coloured spots having diffused borders. It is therefore argued that as a result of metamorphism undergone by these rocks, the coloured spots (pieces of rocks) got slowly absorbed into the surrounding parts of the rocks, and with the completion of the act of absorption, the entire rock (sandstone) acquired a uniform gray or a pink colour (Fig. 2.9 and frontispiece photo).

(f) Some other structures

Sedimentary rocks often times display tracks, trails, casts, moulds, oolites, vugs, concretions, nodules, and so on. The tracks, the trails and the casts are attributed to the activity of organisms, while oolites, pisolites, concretions, vugs and nodules are the outcome of chemical action on the vulnerable rocks like the limestones and the dolomites.

SUMMARY

Below noted features are noteworthy :

1. Different textures, structures encountered are controlled by the actual grain size, the disposition of grains, and the colour of the grains.
2. Considerable variation in the thicknesses of beds, from a few millimetres to a metre or so, is displayed by the rocks.
3. The beds are produced due to a variation in grain size, and/or a pause in the settling down of the grains.
4. Coloured grains produce banded structures.
5. Structures predominate over the textures. Dominant structures are thick/ thin beds, current bedding, graded bedding, ripple marks.
6. Spotted structures constitute an unusual feature.
7. Based on size and shapes of grains, structures/textures recognised are coarse/medium/fine grained rocks, rudaceous/gritty/lateritic structures.
8. Depth of water in the basin, currents produced in the water, and the differential rates of settling of the different grains (sediments) control the development of different structures.
9. Activity of organisms and chemical action produce trails, tracks, vugs, concretions etc.

CHAPTER 3

Petrographical Features of Sedimentary Rocks

INTRODUCTION

The mineralogical composition of the sedimentary rocks is quite simple, because these are almost monomineralic in nature. These rocks are produced through the disintegration and the decomposition of the pre-existing crustal rocks. Further, this process takes place at the surficial temperature and pressure, when the complex silicate minerals of the crustal rocks become unstable, and therefore these break down to simpler oxide combinations. Thus from the pyroxenes, the amphiboles, the felspars and the micas composing the parental igneous and the metamorphic crustal rocks, quartz (SiO_2), clay minerals (SiO_2, Al_2O_3, Fe_2O_3), hydrated iron oxides (goethite, ilmenite) etc., are built up. These products being insoluble in water, these are either left behind at the sites of weathering of the parental rocks, or are transported in suspension by the wind and the running water (streams, rivers). The process of transportation has the potential of separating the products (minerals) of weathering on the basis of inherent specific gravity of the individual minerals. Some of the products of weathering however are soluble ones in water, and these get transported by the rivers and the streams to the basin of sedimentation. These get deposited later as a precipitate. This process gives rise to carbonates, phosphates, sulphates and so on.

The formation of sedimentary minerals through the process of weathering can be understood better from Fig. 3.1 (Pettijohn 1957, Fig. 2.3, p. 106). Study of the said figure reveals that the sedimentary minerals are built from SiO_2, Al_2O_3, Fe_2O_3 (these give rise to sandstones and shales), and MgO, CaO, CO_2 (these give rise to limestones and dolomites). The quartz and the clay minerals are called as the "clastic minerals", while the carbonate minerals are called as the "non-clastic minerals". The clastic minerals are also described as the "detrital minerals". Further distinction is made between "residual detrital" and the "secondary detrital" minerals. The residual detrital minerals are those that survive the process of weathering of the parental rocks. Such minerals are also called as the "relict minerals". The secondary detrital minerals are formed as a result of the breakdown (weathering) of the minerals composing the parental rocks.

Minerals like tourmaline, rutile, zircon, sphene, monazite, titanite etc., occurring as the accessory constituents in the parental igneous and the metamorphic rocks, are resistant to the process of decomposition, and are therefore transported in suspension by the rivers and the streams to the basins. There is yet one more process by which the minerals originate in the sedimentary rocks. It is called as the "authigenic growth" (Figs. 3.2 and 3.3). As a result of the joint operation of these

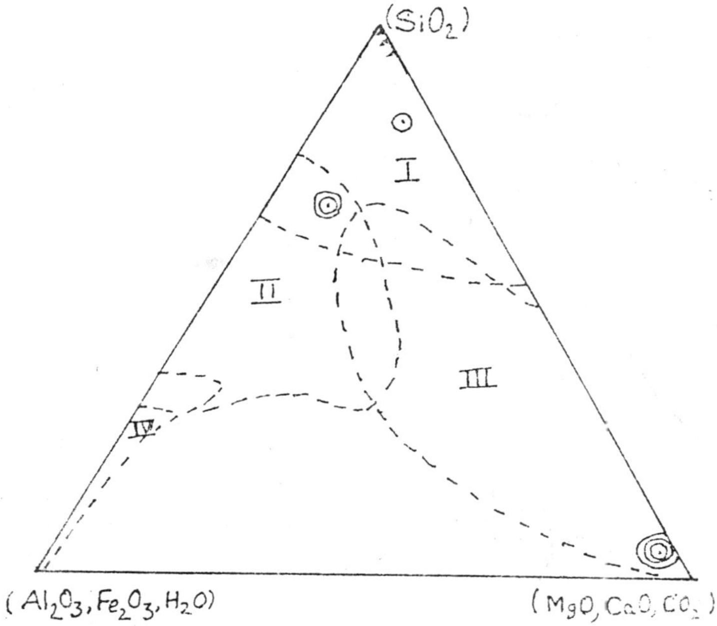

Fig. 3.1. Composition of common sediments (after Fig 42 of Pettijohn, Sedimentary Rocks, 2nd edition, 1957). Positions of "average sandstones, shales, limestones/dolomites and laterites, and other residual accumulations" are indicated by singular or many circles within the respective fields. I. Sandstones, II. Shales, III. Limestones and Dolomites, IV. Laterites and other residual accumulations. Numbers I, II, III and IV have been introduced by the present author.

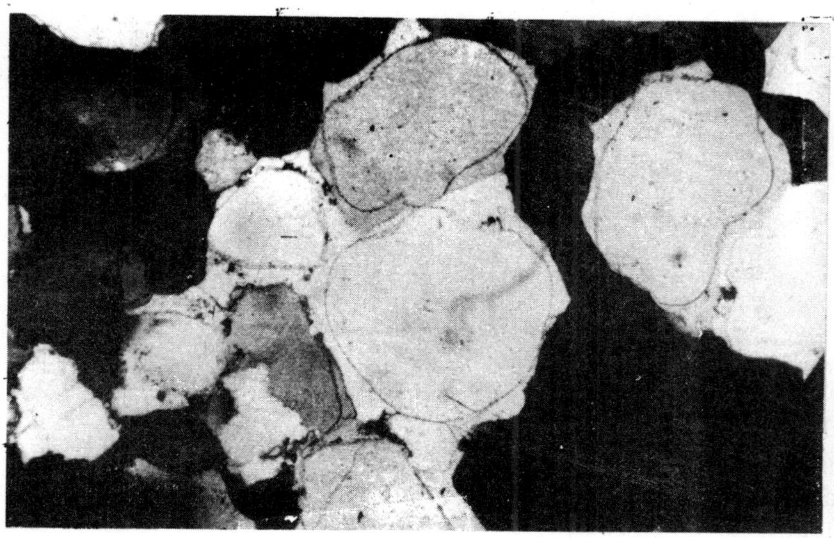

Fig. 3.2. Photomicrograph of quartzarenite showing authigenic overgrowth around detrital grains of quartz. Note the original grain boundary between the detrital grains and the secondary growth around it. Crossed nicols ×3.5. Specimen is from the southeastern part of the Kaladgi basin, Karnataka state. *Courtesy* : Dr. G.V. Hegde.

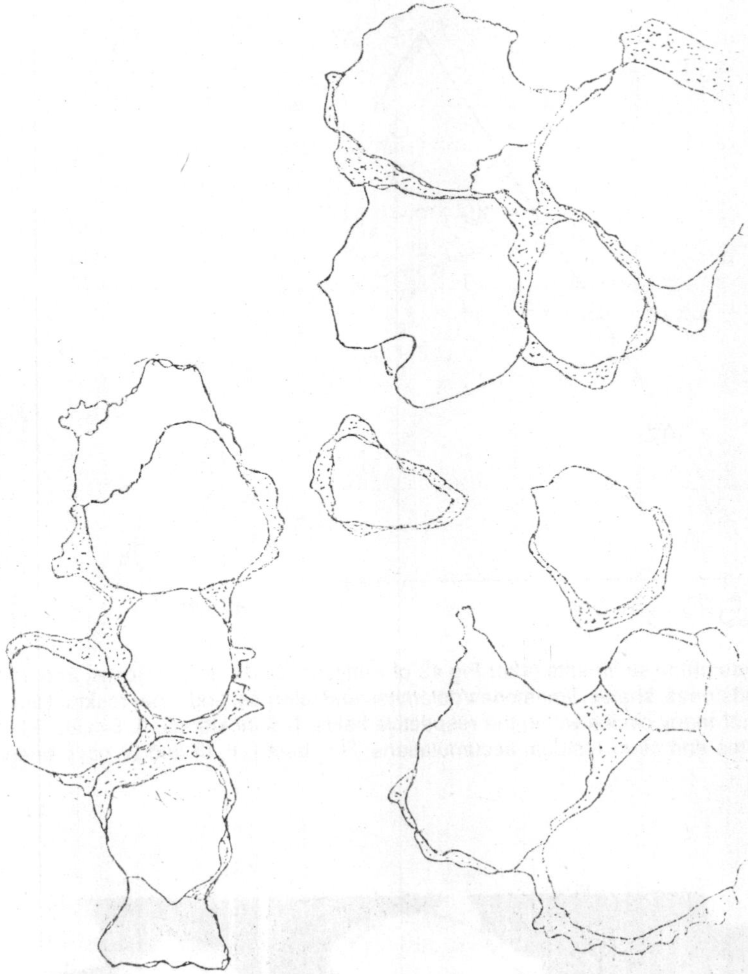

Fig. 3.3. Overgrowth texture. (*Courtesy* : K. Bhimsen. Sample of quartzarenite No. KB 371) [Magnification ×10]. Also see Fig. 3.2.

processes, the sedimentary rocks acquire their mineralogical composition which is indeed unique in several respects.

CLASSIFICATION OF MINERALS COMPOSING SEDIMENTARY ROCKS

The sedimentary rocks themselves are derived from the pre-existing crustal rocks, and are hence classified as the "secondary rocks". Therefore the minerals composing the sedimentary rocks are also considered to be of a "secondary" derivation. The terms "essential and accessory" as are used for minerals occurring in the igneous rocks, cannot be extended to those encountered in the sedimentary rocks, because, majority of these latter rocks are made up of only one or two minerals. As mentioned above, minerals grow around a nucleus of minerals (Figs. 3.2 and 3.3) already occurring in the sedimentary rocks, and to such minerals, the name of "authigenic minerals" is given. Majority of the sedimentary minerals are light in weight, like quartz in a sandstone, kaolinite, montmorillonite, illite etc. in clays and shales. Therefore denser minerals like tourmaline, sphene, titanite etc., which are encountered in the sedimentary rocks are called as the "heavy minerals". Sometimes the pre-existing minerals of the crustal rocks, are not decomposed during the process of weathering, and are transported and deposited in the basin, alongside with the other sediments. Thus grains of felspars are met with, in some sandstones. Such grains (minerals) are thus acquired from the pre-existing rocks, and are hence called as the "relict minerals". The classification of the sedimentary minerals has been presented in Table 3.1.

Table 3.1. Classification of minerals of sedimentary rocks based on the mode of derivation

	Clastic minerals		Non-clastic minerals (precipitates)	Heavy minerals (sp. gr. > 2.85)
Process of formation of minerals	Residual detrital (relict) minerals	Secondary detrital minerals	Products of weathering transported in solution and later precipitated	Obtained from parental crustal rocks, and then transported in suspension to the basin
	Minerals surviving weathering and transported mechanically to the basin	Products of weathering transported mechanically in suspension to the basin		
Names of minerals	Quartz and felspars	Quartz, clay—minerals and so on	Calcite, dolomite, aragonite, siderite, and so on	Zircon, monazite, sphene, rutile, and so on

The sedimentary rocks are not only made up of minerals, but even pieces of parental rocks are found in them. In this respect it may be said that the composition of the sedimentary rocks is quite heterogeneous one. The point to be considered is that the crustal rocks get decomposed, and some of the products are left behind at the site of decomposition. Such products also are to be considered as rocks, even though these are not compacted into hard beds or layers. Hence the soils (which are the important products of weathering) are required to be categorised as rocks. Soil is a residual formation which is not transported into any of the basins of sedimentation. There is a variety of sedimentary rock called as a "graywacke" which contains pieces of rocks cemented in a matrix. Mineralogically, composition of such a rock is very heterogeneous. Another feature to be noted is that the sedimentary rocks possess a matrix that holds the grains (minerals) together. The matrix might consist of very fine grained minerals. Therefore the minerals encountered in the sedimentary rocks are considerably varied ones. Generally the minerals composing the matrix are not taken into account while describing the mineralogical composition of the sedimentary rocks.

CLASSIFICATION OF SEDIMENTARY ROCKS

Several bases are considered in the classification of these rocks, the bases adopted being (i) mineralogical composition, (ii) the nature of sediments, (iii) the chemical composition, and (iv) the colour and the grain size. Some of the bases give rise to a classification that can be readily adopted during the field studies, like the colour and the grain size. Others can be made use of only in the laboratory, like the basis of chemical composition. Thus the classifications serve different purposes like making distinction between the several varieties, establishing origin of the rocks and so on. There is one basis which is exclusively applicable to the sedimentary rocks. It is the presence of fossils in some of them. Thus rocks are named as the foraminiferal limestones, the crinoidal limestones and so on. Classifications of the sedimentary rocks based on the above noted bases are described below.

(a) Classification based on the mineralogical composition

This is the most important basis, because, by definition, a rock is an aggregate of minerals, and therefore, minerals must be considered first, in effecting classification of any rock. As many as 51 minerals are encountered in the sedimentary rocks, and these comprise of light minerals (sp. gravity less than 2.85), and of heavy minerals (sp. gravity more than 2.85). Frequency of occurrence of them is not the same. It has been observed that only 7 minerals belong to the class of light minerals, while 44 form the category of heavy minerals. It is further interesting to note that though heavy minerals are more in number, these according to Pettijohn constitute a mere one per cent of the arenaceous sediments (Pettijohn, 1957, Table 24, p. 117). Further, only 20 heavy minerals are commonly met with, while 24 are not frequently encountered in the sedimentary rocks. Thus it is seen that though the heavy minerals are more in number, their participation in the building up of the sedimentary rocks is negligible. Classification of the sedimentary minerals on the basis of frequency of occurrence is set out in Table 3.2.

Table 3.2. Classification of sedimentary minerals based on frequency of occurrence and their sp. gravity (slightly modified from Table 24 of Pettijohn, 1957, p. 117)

Most common minerals		Heavy minerals (sp. gravity > 2.85) occasionally met with. These constitute less than one per cent of the rock
Light minerals (sp. gravity < 2.85). All occur very frequently excepting chalcedony	Not very frequently encountered (sp. gravity > 2.85)	
Quartz, felspars, calcite, dolomite, clay minerals, chalcedony (not very common)	Biotite, garnet, hornblende, hypersthene-enstatite, ilmenite, chlorite, magnetite, rutile, sphene, tourmaline, zircon, zoisite, staurolite, leucoxene, kyanite, epidote, augite, diopside, apatite, andalusite	Antinolite-tremolite, anatase, barite, brookite, casseterite, chloritoid, clinozoisite, collophane, cordierite, corumdum, dumortierite, fluorite, glaucophane, hematite, limonite, monazite, olivine, pyrite, siderite, sillimanite, spinel, topaz, xenotime, serpentine

It will be apparent from Table 3.2 that the most frequently occurring minerals in the sedimentary rocks are quartz, felspars, clay minerals, calcite and dolomite. A sandstone is made up practically entirely of quartz, a shale is composed predominantly of clay minerals, while calcite together with small quantity of dolomite, build up the limestones. The sedimentary rocks therefore could be described as having a "monomineralic composition". Hence the sedimentary rocks are classified considering the above noted frequently occurring minerals and the same is presented in Table 3.3.

Table 3.3. Classification of sedimentary rocks based on mineralogical composition

Predominant minerals	Other minerals	Rock name
Quartz	Sericite, felspars, and minerals forming matrix	Sandstone
Quartz (predominant) plus felspars	As above	Felspathic sanstone or arkose
Clay minerals (predominant) plus quartz, limonite, goethite	Minerals forming matrix	Shale, argillite, mudstone, clay beds
Calcite, dolomite	Some matrix minerals, aragonite, siderite	Limestone, dolomitic limestone

Rocks noted in Table 3.3 may contain heavy minerals like pyrite, zircon, apatite, tourmaline, sphene and so on, but their presence or absence does not alter the rock name.

From the foregoing account it may be said that the mineralogical composition of the sedimentary rocks is quite simple, in the sense that only one or two minerals share the entire composition of them.

(b) Classification based on the character of the sediments

Actually the composition of the sedimentary rocks is controlled by that of the sediments reaching the basin. The composition of the sediments, in turn, depends upon the distance of transport undergone by the products of weathering. The intensity (distance) of transport reduces the grain size, and brings about a separation of the soluble from the insoluble products of weathering. Considering the actual size of the particles (grains) composing the sediments, several families of sediments are recognised. Names are given to the families, and to the rocks formed from them. Table 3.4 sets out the details.

Table 3.4. Classification of sedimentary rocks based on character of sediments

Grain (particle) size (in mm)	Character of sediment	Family name	Rock name
2 to 256	Rudaceous (boulders, cobbles, pebbles, granules)	Rudites	Boulder bed conglomerate
1/16 to 2	Arenaceous (sand)	Arenites	Sandstones
1/16 to 1/256	Argillaceous (silt)	Argillites	Argillites
Less than 1/256	Clay	Clay	Clay, shale, mudstone

It will be appreciated from Table 3.4 that the actual minerals composing the rocks are not taken into account. Therefore this basis is quite useful in effecting classification during the field studies through the megascopic estimation of the grain size of the rocks. Such a classification is readily adoptable, and it helps merely to identify the different families amongst the sedimentary rocks. This scheme, however, cannot be extended to the carbonate rocks (limestones and dolomites), because carbonate materials are transported in solution, and not in a state of suspension, as are the rudaceous, the arenaceous and the argillaceous materials.

(c) Classification based on chemical composition

Chemical analysis becomes necessary when the grain size becomes too fine for identification of minerals. In order to appreciate the chemical differences between the three dominant sedimentary rocks namely the sandstone, the shale and the limestone, their chemical composition is set out in Table 3.5.

Table 3.5. Chemical composition (average values) of sandstone, shale and limestone (after Pettijohn, 1957, Table 20, p. 106)

Oxide (in per cent)	Sandstone	Shale	Limestone
SiO_2	78.33	58.10	5.19
TiO_2	0.25	0.65	0.06
Al_2O_3	4.77	15.40	0.81
Fe_2O_3	1.07	4.02	0.54
FeO	0.30	2.45	—
MgO	1.16	2.44	7.89
CaO	5.50	3.11	42.57
Na_2O	0.45	1.30	0.05
K_2O	1.31	3.24	0.33
H_2O	1.63	5.00	0.77
P_2O_5	0.08	0.17	0.04
CO_2	5.03	2.63	41.54
SO_3	0.07	0.64	0.05
BaO	0.05	0.05	—
C	—	0.80	—

Scrutiny of Table 3.5 reveals that only 5 oxides share the chemical composition of the three dominant sedimentary rocks, and these are set out in Table 3.6 in order to bring out the differences between the said three dominant rocks.

Table 3.6. Diagnostic oxides of the dominant sedimentary rocks

Oxide (in per cent)	Sandstone	Shale	Limestone
SiO_2	78.33	58.10	5.19
Al_2O_3	4.77	15.40	0.81
CaO	5.50	3.11	42.57
MgO	1.16	2.44	7.89
CO_2	5.03	2.63	41.54

The study of chemical features set out in Table 3.6 helps to classify the sedimentary rocks into three major groups namely :

1. Highly siliceous group.
2. Moderately siliceous together with considerable quantity of alumina, group.
3. Highly calcareous group.

Considering the chemical composition, the sedimentary rocks get classified into sandstones (highly siliceous group), shale (predominance of alumina group), and limestone (highly calcareous group). The basis of chemical composition assists in identifying very fine grained sedimentary rocks with a fair degree of accuracy, if mineral identification be not possible. Goldschmidt (1937) has classified the sedimentary rocks on the basis of the contents of the major oxides into five categories namely :

1. Resistates (SiO_2).
2. Hydrolysates (clay minerals, Al · Si).
3. Oxidates ($FeOH_3$).
4. Carbonates, precipitates [$CaCO_3$, $CaMg(CO_3)_2$].
5. Evaporates ($NaCl$, KCl, $MgCl_2$, $MgSO_4$, etc.).

The resistates stand for the sandstones, the hydrolysates for the shales, the carbonates-precipitates for the limestones and the dolomites. It is thus seen that according to Goldschmidt once again, only three rocks are recognisable. In addition he distinguishes a laterite. Very rarely evaporites give rise to salt beds, like the Saline series of Salt Range, which is more than 450 metres thick, the Stassfurt deposits of Germany and so on.

(d) Classification based on the size and the shape of grains

Consideration of the actual sizes of grains gives rise to the coarse, medium and the fine grained rocks, but it does not introduce any new variety of sedimentary rock. Generally the arenaceous rocks are coarse to medium in grain size, while the argillaceous rocks are very fine grained in nature. The calcareous rocks are not distinguished on the basis of the grain size, though these are also fine grained in nature, like the argillaceous rocks. The limits set for the coarse, the medium and the fine grain sizes have already been given in Chapter 2.

On the basis of the shapes of the grains, more varieties are recognised. If the grains be big enough and these be rounded to sub-rounded in shape, these produce a conglomerate. If the grains be small in size, but are numerous, these give rise to a grit (Fig. 2.14). In case the grains be angular or irregular in shape, these produce a breccia (Figs. 2.15a, b, c). Sedimentary rocks classified by a combination of size and shape of grains, is presented in Table 3.7.

Table 3.7. Classification of sedimentary rocks based on the combination of size and shape of grains

Size (in mm)	Shape	Name of rock
Very, very coarse (256 to 512 and above)	Giant sized boulders	Boulder bed
Very coarse (64 to 256)	Rounded to sub-rounded pebbles, cobbles	Conglomerate
Coarse to medium (4 to 64)	Angular sub-angular, and irregular	Breccia
Medium to fine (2 to 16)	Nearly regular in shape like rectangular, cubic and numerous in number	Grit
Coarse (2 to 1/2)	Nearly rounded	Coarse grained sandstone
Medium (1/2 to 1/4)	Nearly rounded	Medium grained sandstone
Fine (1/4 to 1/16)	Nearly rounded	Fine grained sandstone
Very fine (1/16 to 1/1024)	Not detectable	Shale, argillite, clay rock
Very fine (1/16 to 1/1024)	Not detectable	Limestone, dolomite

(e) Classification based on other bases

Colour, external appearance, type of fossils, environment of sedimentation are also made use of, in classifying the sedimentary rocks. But these bases do not add to the rocks recognised on the basis of mineralogical composition, and the shapes and sizes of the grains. These are used as prefixes or as adjectives to the rock names (sandstone, shale etc.).

Colours encountered with the sedimentary rocks are red, pink, violet, buff, gray and so on. As already mentioned, the Red Fort of Delhi is built of red coloured sandstones. The World famous Taj Mahal of Agra is constructed from white coloured marble occurring at Makrana, Rajasthan, which prior to its getting metamorphosed to marble, was a limestone. Shales display a greater variety of colours. Some of the sedimentary rocks contain rock fragments to a considerable extent. Therefore such rocks cannot be classified only on the basis of the minerals or the grain size. Especially some of the sandstones are observed to contain fragments of slate, phyllite, siltstone, quartzite and so on. The grains of quartz and of felspar are found to be extremely angular, and there is a preponderance (around 15%) of matrix. Such a rock is classified as a graywacke. This rock has a considerable resemblance to a breccia. In fact Pettijohn (1957, p. 303) has written about the graywacke as "the term breccia is most appropriate".

The structure and the external appearance is sometimes made use of, in classifying some rocks. The oxides of iron, manganese and alumina being the insoluble products of weathering, these form a residual mass at the site of weathering. It possesses a porous, concretionary structure (Fig. 3.4), and has a reddish brown colour. Such a product is classified as a laterite. Grain size analysis not only helps to recognise very coarse, coarse, medium, fine and very fine grained sedimentary rocks, but it further serves as a basis to classify the arenaceous rocks in more details. While describing the textures and the structures of the sedimentary rocks (Chapter 2), mention has been made to the class-limits set by Wentworth (1922). It is customary to determine the sizes of the grains composing loose sediments by subjecting the samples to the sieve analysis. In the case of rocks, the mean grain sizes are determined from their thin section, a detailed account of which is given in Chapter 7. The frequency percentage values of mean grain sizes are plotted against the class-limits of Wentworth. Such a plotting gives rise to curves with one, two or more number of peaks (Figs. 3.5a, b, c, d). Development of such peaks helps to recognise "unimodal, bimodal, and polymodal or irregular" varieties of arenaceous rocks. The suffix "modal" does not stand for the contents of the minerals, but to the development of peaks in the frequency distribution curves. This procedure helps to recognise homogeneous and heterogeneous textures in the analysed rocks. This has a resemblance to the textural varieties like the equigranular, inequigranular textures encountered in the igneous rocks. Unimodal size distribution compares well with the equigranular texture of the igneous rocks. Unimodal texture may be further split into very coarse, coarse, medium and fine varieties according to the location of the singular peak in the class-limits marked on the abscissa. When two peaks are produced, it gives rise to a bimodal variety, which is comparable to the porphyritic texture of the igneous rocks. When more than two peaks are produced, it gives rise to the "irregular or polymodal" type, which may be compared to the "seriate" texture of the igneous rocks. The bimodal and the polymodal textures are indicative of the grain size being of a heterogeneous nature. The "modal basis" thus assists in recognising sub-classes within the sandstones such as (i) coarse grained, (ii) medium grained, (iii) fine grained, (iv) equigranular, (v) inequigranular, (vi) heterogeneous grained and so on.

Fig. 3.4. Field photo of a laterite exposed in a road cutting near Hunerveli, Shrivardhan, Kolaba district, Maharashtra state. Note the characteristic porous structure which is called as the "lateritic structure". *Courtesy :* Dr. A.H. Kouhsari.

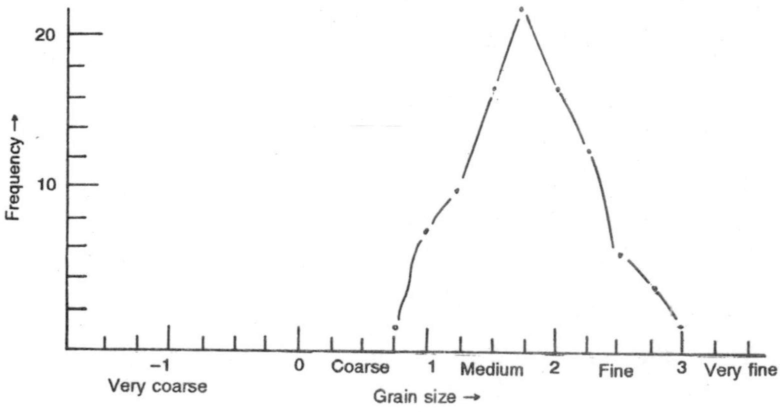

(a) Sample No. PM 119 of quartzarenite. *Courtesy :* G.S. Pujar, 1989.

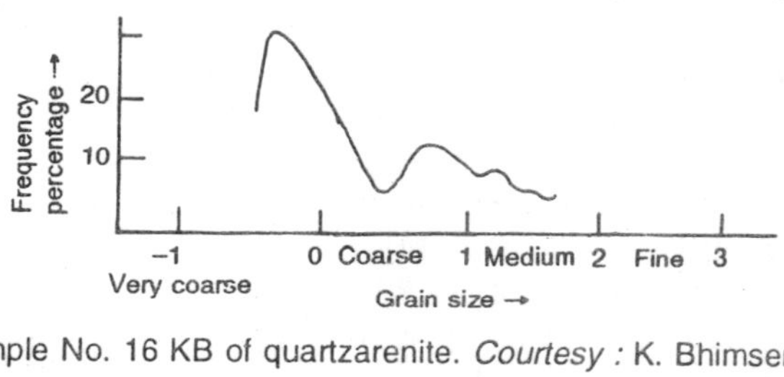

(b) Sample No. 16 KB of quartzarenite. *Courtesy :* K. Bhimsen, 1989.

Figs. 3.5a, b. Frequency distribution curves of grain size displaying unimodal pattern. (a) Peak developed in medium-fine grain size. (b) Peak developed in very coarse grain size.

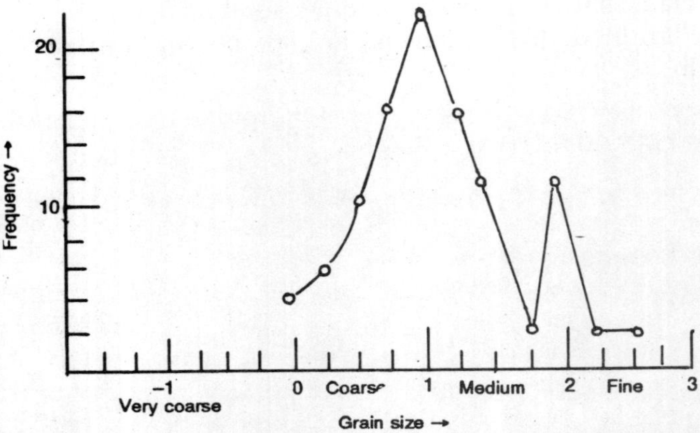

(c) Sample No. PM 61 of quartzarenite. *Courtesy* : G.S. Pujar, 1989.

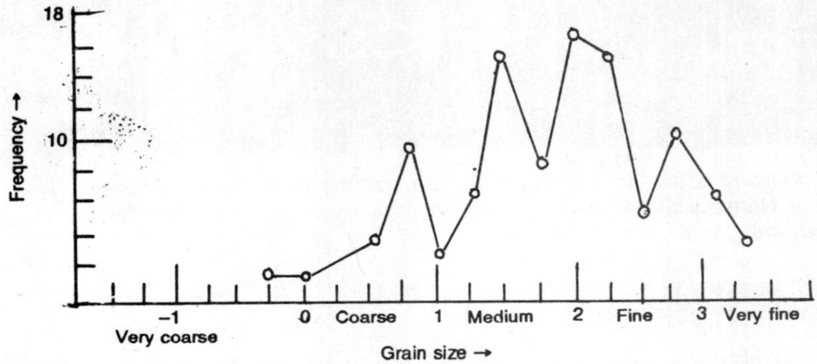

(d) Sample No. PM 53 of quartzarenite. *Courtesy* : G.S. Pujar, 1989.

Figs. 3.5c, d. Frequency distribution curves of grain size. (c) Bimodal distribution pattern with peaks in coarse-medium and medium-fine grain sizes. (d) Polymodal or irregular distribution pattern with peaks in different grain sizes. Coarse-medium, medium, medium-fine and fine-very fine categories.

The basis of environment of formation of the sedimentary rocks is more useful while commenting on the origin of the rocks. At this juncture the recognition of "magmatic, metamorphic-metasomatic and the anatectic" categories of granites is very relevant. The said distinction of the granites is deemed as a classification of granites. Turning back to the sedimentary rocks, the environments under which the sediments are laid (and hence formed) are "aeolian, beach, shallow marine, fluvial and turbidity". Procedure of their establishment has been described in Chapter 7 which deals with the topic of Sedimentary Analysis. Adopting such a technique, it is possible to classify sandstones into different environments according to the nature of the grains composing the rocks. However the sedimentary analysis is applied only to the arenaceous rocks (sandstones). Therefore the environment (aeolian etc.) derived for the sandstones is also valid for the associated shales, limestones and so on. It is necessary that all these rocks belong to one and the same basin, and that they are formed during one and the same period.

From the foregoing account of the classification of the sedimentary rocks based on the different bases, it is quite evident that the rocks are few in number, and their mineralogical composition is also very simple. There are only 9 rocks that differ considerably from one another. These are sandstone, shale argillite, limestone, dolomite, conglomerate, grit, breccia, and laterite. Sub-divisions are made in each rock based on grain size, colour, environment of formation and so on.

NOMENCLATURE OF SEDIMENTARY ROCKS

It is customary to name the rocks on the quality and quantity of minerals present in them. The geographic location of the place where the rock was first discovered, external appearance of the rock are also utilised in naming them. Majority of names of the sedimentary rocks end with the suffix "ite" like, laterite, dolomite. Sedimentary rocks are mostly composed of one or two minerals. Such rocks are therefore named after the predominant minerals like sandstone (quartz), limestone (calcite), mudstone (clay minerals). The name laterite is so given owing to the inherent property of yielding "dimension stones" like bricks. The word "later" means a brick, and a rock which yields such stones has been rightly named as a laterite. Rock name graywacke is derived from the distinctive dark gray colour of the fresh rock, the name being given by Krynine and Folk (Pettijohn, 1984, p. 212). Only the sedimentary rocks are named on the basis of the grain size. Thus the names rudite, arenite and argillite are given exclusively on the sizes of the grains, but these are family names, and are not rock names. Flagstones are in fact sandstones which are thinly laminated. Such rocks yield thin slabs or flags. The name flagstone therefore is based on its physical property.

Thus it may be said that the nomenclature of the sedimentary rocks is by and large quite simple and the names readily convey the mineralogical composition.

DESCRIPTION OF SEDIMENTARY ROCKS

The major groups or classes recognised as a result of classifying the sedimentary rocks will be described in respect of the mineralogical composition, and the textural/structural features. Each group is sub-divided in order to understand the inter-relations between the members, and their origin. A schematic classification of the sedimentary rocks is given in Fig. 3.6.

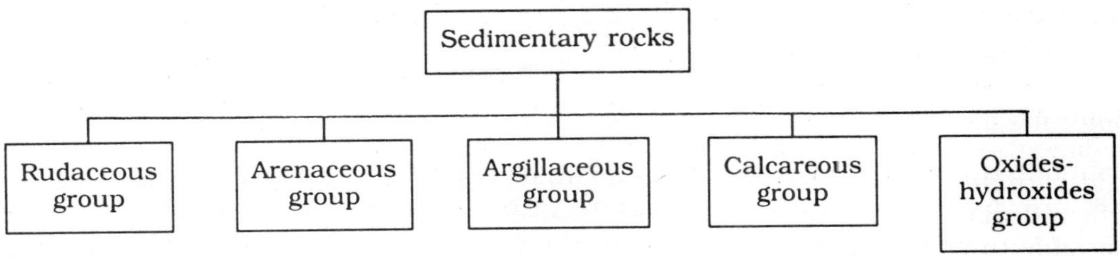

Fig. 3.6. Schematic classification of sedimentary rocks.

A detailed account of each group of sedimentary rocks is given, in the following pages.

(a) Rudaceous group

Rocks belonging to this group are characterised by the presence of extremely large sized constituents (grains) in them. Not only the sizes are varied, but the composition of the constituents is also varied one. In fact, the ingredients are pieces of rocks, in the great majority of cases, but in some instances, minerals do build the rudaceous rocks. Rock pieces or the mineral grains are cemented together in a matrix giving rise to a conglomerate, a grit, or a breccia. The rock pieces or the minerals might be rounded to sub-rounded, or angular to sub-angular in shape. Depending upon their actual sizes, these are termed as a boulder, a cobble, a pebble or a granule (Table 2.2). The family name for the rudaceous rocks is "rudites", Greek word for which is "psephites". These rocks are also described as "gravelly rocks". The classification of the "Rudaceous Group" and of "Rudites/Psephites" is presented in Fig. 3.7.

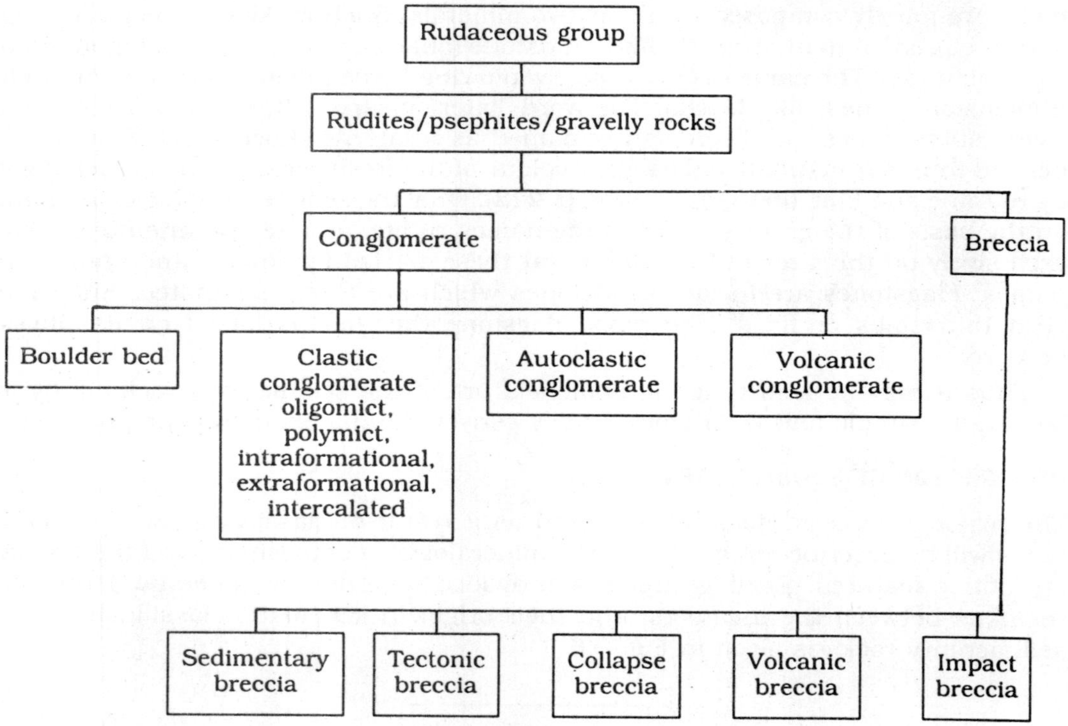

Fig. 3.7. Schematic classification of Rudaceous group and of Rudites/Psephites.

Conglomerate

It is made up of cobbles, pebbles, boulders or granules of rocks or of minerals, and these are cemented in a matrix, which might be siliceous, argillaceous or calcareous one. The boulders, the pebbles etc., could be rounded to sub-rounded in shape (Figs. 3.8a, b, c). At times these might be even lenticular or elongated in form. These rocks are generally found at the bottom, or the basal parts of a sedimentary basin. However conglomerates might occur at a higher horizon in a sedimentary succession (Figs. 2.10 and 3.9). Both in a gravel bed or in a conglomerate, the pebbles, the cobbles are unassorted. The term conglomerate is applied only when the gravels are indurated (compacted).

The conglomerates are divided into 4 sub-divisions (Fig. 3.7) namely (i) boulder bed, (ii) clastic, (iii) autoclastic, and (iv) volcanic. The clastic conglomerates are

3.8 (a)

3.8 (b)

Contd.

Fig. 3.8a, b, c. Field photos of conglomerate beds encountered in the vicinity of Daddi village (Belgaum district, Karnataka state), on the banks of Ghataprabha river. Note the rounded to subrounded shapes of the pebbles as well as the smaller sized (as compared to that seen in Fig. 2.8a, which is a boulder bed) pebbles. At 3.8b, the conglomerate is observed to rest over a bed of quartzarenite. At 3.8c, a diversity in the sizes of the pebbles and the boulders can be derived from the nature of the cavities left by them. *Courtesy :* Dr. K. Bhimsen.

Fig. 3.9. Field photo of an intercalated conglomerate layer occurring in between two beds of pink quartzarenite. The conglomerate layer is about 20-30 cm thick. The exposure is located 1 km west of Vatnal village, Belgaum district, Karnataka state. *Courtesy :* Dr. G.S. Pujar. (In Fig. 2.10, two layers of intercalated conglomerate are observable).

further split into 5 types namely, (a) oligomict, (b) polymict, (c) intraformational, (d) extraformational and (e) intercalated (Fig. 3.7). The term clastic indicates that the pebbles, the cobbles etc., were transported to the site of sedimentation by the rivers, the streams, and the glaciers. Such a distinction is necessary because, some conglomerates are formed in situ (tectonic or volcanic activity).

The terms "oligomict and polymict" have been introduced by Schwetzoff (1934) into the description of the conglomerates. Oligo means few, poly means many, and mict means mixed. These terms (oligo, poly, mict) are derived from Greek words (Pettijohn 1957, p. 255). The oligomict conglomerate is also called as "orthoquartzitic conglomerate" which is composed of only one kind of rock pieces, or minerals, like quartz, quartzite, chert or a mixture thereof. The grain size is not coarse. The polymict conglomerate is also called as "petromict conglomerate". The pebbles, and the cobbles are of igneous, sedimentary or metamorphic rocks. The presence of a variety of (in the sense of composition) pebbles and cobbles, is the salient feature of this variety, and is therefore called as the "polymict conglomerate". The Bagewadi-Bannikoppa conglomerate from the Gadag Schist Belt, belongs to such a category, as has been described by Gokhale et al. (1971). A conglomerate which derives its fragments (pebbles etc.) from within the basin of deposition is called as an "intraformational conglomerate". When the pebbles, cobbles etc., are brought from outside the basin, such a rock is classified as an "extraformational conglomerate". This is the most common variety. It is further split into 2 sub-varieties. If the conglomerate occurs at the base of the sedimentary succession, it is called as a basal conglomerate. On the other hand, if it be encountered at a higher horizon, it is designated as an "intercalated conglomerate" (Figs. 2.10 and 3.9).

The autoclastic conglomerate is characterised by the presence of pebbles of only one rock like a quartzite, a granite, a banded hematite quartzite and so on. In fact such a conglomerate is produced in situ through the fracturing and fragmentation of the pre-existing rocks by the action of the dislocatory movements like faulting. Boudinage (lensoid) structure developed on the limbs of the folds during the process of folding of rocks can also lead to the formation of the autoclastic conglomerate. The term autoclastic means a material being rendered sedimentary like (clastic) without the action of weathering or any transportation by the geological agents. The banded hematite quartzites exposed in the north-eastern vicinity of the Jambunath Temple, Jambunath hills, Bellary district, Karnataka state, have developed autoclastic conglomerate (Fig. 3.10). Quartzarenitic rocks exposed near Yekkeri, Belgaum district, Karnataka state, have also produced such a rock (Fig. 3.11). Hayden (1909) has described occurrence of autoclastic conglomerate layers amongst the slates of Cambrian of Spiti area. It may be said that the autoclastic conglomerate somewhat resembles an "intraformational conglomerate" because the former variety is also developed "in situ" i.e., in place where the rock is now found to occur.

The name "boulder bed" is applied to a conglomerate which is composed of very large sized blocks of the sizes of boulders (Fig. 3.12a, b). However the pebbles, the boulders etc., are rounded to sub-rounded in shape. Maclaren (1906) has described such an occurrence from the Bagewadi-Bannikoppa area, Gadag Schist belt, Dharwad district, Karnataka state. Other examples are the Blaini boulder bed, the Tanakki boulder bed of Kashmir-Hazara, the Mandhali beds of Tehri-Garhwal, the boulder beds of Lachi as described by Krishnan (1968). Fundamentally, a boulder bed is a conglomerate. Volcanic activity also can give rise to a conglomerate. Roy and

Fig. 3.10. Field photo documenting very well developed autoclastic conglomerate from banded hematite quartzite. Note the ellipsoidal shapes of the boulders and the pebbles. Also note the plunge of the longer axes of the ellipsoidal pebbles/boulders (handle of the hammer indicates the plunge of the longer axes). Extreme diversity of the sizes of the pebbles is a salient feature. Whereas in Figs. 2.8a, b and 3.8a, b, c (these are also conglomerate beds) the pebbles display typical rounded to subrounded shapes. The conglomerate documented in Fig. 3.10 displays pebbles having ellipsoidal shapes. The autoclastic conglomerate is exposed in the immediate northeastern vicinity of Jambunath temple, Hospet, Bellary district, Karnataka state. *Courtesy :* Dr. H.D. Desai.

Fig. 3.11. Field photo of excellently developed autoclastic conglomerate from the quartzarenitic rocks. Note the typical ellipsoidal shapes and the diversity of the sizes of the pebbles and the boulders. Hammer is by the side of the biggest ellipsoidal boulder measuring 90 × 15 cm. This autoclastic conglomerate is exposed at a distance of 3.5 km east of Yekkeri village, Belgaum district, Karnataka state. *Courtesy :* Dr. G.S. Pujar.

3.12 (a)

3.12 (b)

Fig. 3.12a, b. Excellently developed boulder bed in the vicinity of Bannikoppa and Bagewadi villages, Shirhatti taluka, Dharwad district, Karnataka state. Note the enormous sizes of the boulders, but at the same time these are completely rounded in shape. Composition of the boulders is typically varied one. Thus boulders of granite, basic dyke and others, are encountered in this boulder bed. *Courtesy :* Dr. V.B. Koppad (Fig. 3.12a) and Dr. N.N. Gothe (Fig. 3.12b). Fig. 2.8a, also is an example of a boulder bed.

Biswas (1982) have described volcanic conglomerate/pyroclast horizon near Palkanmaradi in the Hutti schist belt.

Heterogeneous composition and a considerable variation of sizes of pebbles, cobbles etc. is the most salient aspect of the conglomerates. Whenever diverse ideas or views are mixed up, causing confusion, it is customary to describe it by the expression "conglomeration of things". Unassortedness is the common factor between the rock "conglomerate" and its use in "English literature" through the adjective "conglomeration".

Breccia

This rock is composed of angular grains of fragments of rock and these are set in a matrix of similar composition like that of the fragments, but these are of a finer grain size. In other words it means that the fragments and the matrix are of identical composition. Breccia is divided into 5 varieties namely (i) sedimentary, (ii) tectonic, (iii) collapse, (iv) volcanic, and (v) impact varieties. A sedimentary breccia is one which is formed during the process of sedimentation in the basin, when other sediments like sands, clays are laid. Probably angularities of the grains is due to a short distance of transport of the sediments. A tectonic or a cataclastic breccia is formed along a fault plane, owing to the differential movement taking place along the fault plane. Hegde (1987) and Pujar (1989) have described instances of fault breccia. There is another variety which is produced due to the collapse of rocks owing to lack of support from below. Generally a collapse breccia is associated with limestones overlain by resistant rocks like sandstone, quartzites and so on. Limestone gets leached away by the chemical action as a result of which the overlying rocks are rendered supportless, and hence buckle downwards resulting in a jumbled mass of angular rock fragments of different sizes and shapes. During violent eruption of lava from the crater, pieces of rocks are torn apart from the wall of the vent and these are thrown into the air. Such pyroclastic pieces fall down and get cemented together in a fine grained to glassy matrix. This rock is called as a "volcanic breccia". Ash beds of the upper parts of the Deccan Traps around Bombay and Pune reveal a brecciated structure, fragments of traps being found in a matrix of dusty or fine grained material (Krishnan, 1968, p. 406). These are instances of volcanic breccia. Impact breccia is one which is formed by the impact of meteorites on the surficial rocks of the earth. Such an impact can bring about shattering of the crustal rocks into pieces which later get cemented in a fine grained matrix.

(b) Arenaceous group

This is by far the most frequently developed group amidst all of the sedimentary rocks. Such of the rocks that consist of grains varying between sizes of 2 and 1/16 mm get the family name of arenites. These rocks are commonly known as the sandstones, because the grains varying between sizes of 2 and 1/16 mm are technically called as sand. Even though the grains of quartz dominate in the sandstones, those of felspars are also encountered. Therefore the arenites are classified on the nature and the contents of the minerals occurring in them. Three parameters namely :

(i) the source index or the ratio of felspars to the rock fragments,

(ii). the maturity index or the ratio of "quartz (inclusive of chert) to the felspars plus the rock fragments", and

(iii) the fluidity index or the ratio of sand to the matrix, are also considered in the classification of the arenites.

A tabular classification given by Pettijohn (1957) is reproduced in Table 3.8.

Table 3.8. Classification of sandstone (after Pettijohn, 1957, Table 48, p. 291)

	Cement or matrix	Detrital matrix prominent (over 15%) to predominant. No chemical cement	Detrital matrix absent or scanty (under 15%). Voids empty or filled with chemical cement		
Sand or detrital fraction	Felspars exceed rock fragments	Felspathic graywacke	Arkosic sandstones		Orthoquartzites (also called as quartzarenites) with less than 5% chert
			Arkose	Sub-arkose or felspathic sandstone	
Sand or detrital fraction	Rock fragments exceed felspars	Lithic graywacke	Lithic sandstones		Orthoquartzites (also called as quartzarenites) with greater than 5% chert
			Sub-graywacke	Protoquartzite	
	Quartz content	Variable, generally less than 75 per cent	Less than 75 per cent	Between 75 and 95 per cent	More than 95 per cent

It will be appreciated from Table 3.8 that there are only two distinguishable varieties within arenites (commonly known as sandstones) namely, the sandstones and the graywackes, which could be easily identified megascopically in the field studies. Other rocks of the arenaceous group (arenites) require thin section examination in the laboratory. Petrographic features of the different rocks belonging to the arenite family are described in the following pages.

Sandstones

These rocks are more predominant than the graywackes. A sandstone is composed practically entirely of grains of quartz (75 to 95 per cent) that are set in a siliceous, argillaceous, ferruginous or calcareous matrix. The matrix is absent or forms only about 15 per cent of the rock. In some sandstones "polycrystalline grains of quartz" are met with (Figs. 3.13a, b and 3.14). The grains of quartz are characteristically rounded to sub-rounded in shape. Grains of felspars, flakes of mica, heavy minerals (if any), and rock fragments together form around 20 to 25 per cent of the rock, but the content of felspars is nearly equal to that of the rock fragments. Authigenic overgrowth is often encountered (Figs. 3.2 and 3.3) around grains of felspars, quartz, tourmaline and so on. Depending upon the proportion of the felspars and the rock fragments, the sandstones are further split into varieties like the arkosic sandstone and the lithic sandstone. In an arkosic sandstone, felspars are more than the rock fragments. If the content of quartz be around 75 per cent, it is classified as an arkose. If the content of quartz be between 75 and 95 per cent, it is designated as a sub-arkose or as a felspathic sandstone. In a lithic sandstone, rock fragments are

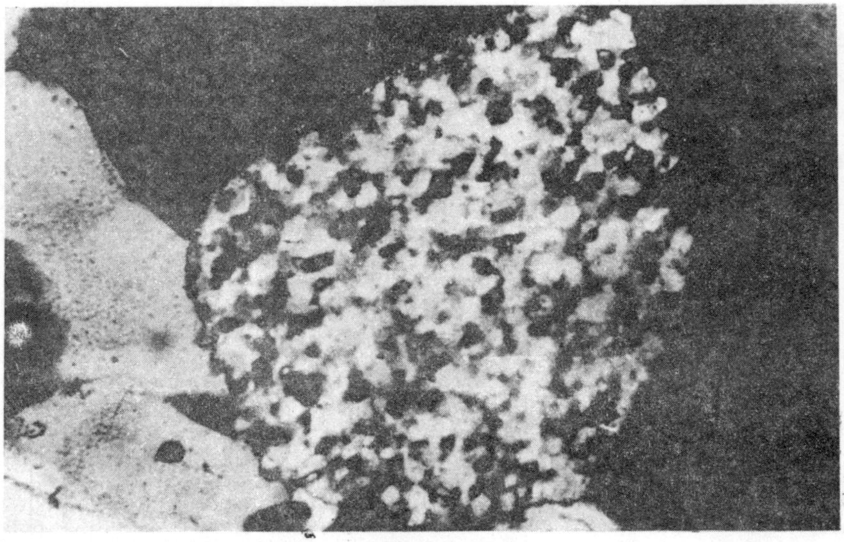

Fig. 3.13a. Photomicrograph of a quartzarenite showing a polycrystalline grain of quartz in it. Note the oval shape of the grain, and sutured borders of the multitudinous small grains forming the oval shaped grain. Crossed nicols ×10. Rock specimen is from the southeastern part of the Kaladgi basin, Karnataka state. *Courtesy* : Dr. G.V. Hegde.

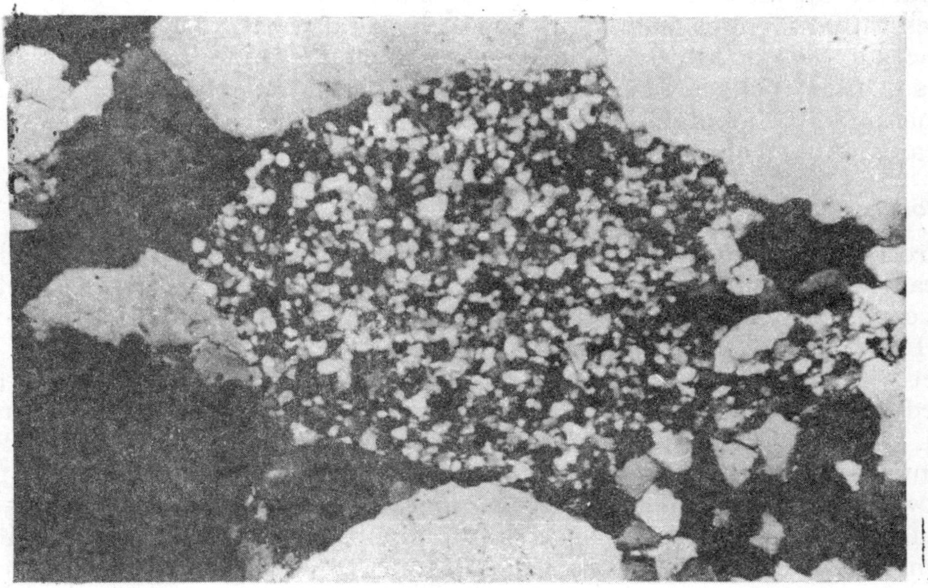

Fig. 3.13b. Photomicrograph of a polycrystalline grain of quartz occurring in a quartzarenite. Note the presence of multitudinous tiny grains of quartz within the framework of one-boundary. As a contrast, note the considerably larger and bigger sizes of the adjacent individual grains. Crossed nicols ×100. The rock specimen is from Manoli village, Belgaum district, Karnataka state. *Courtesy* : Dr. G.S. Pujar.

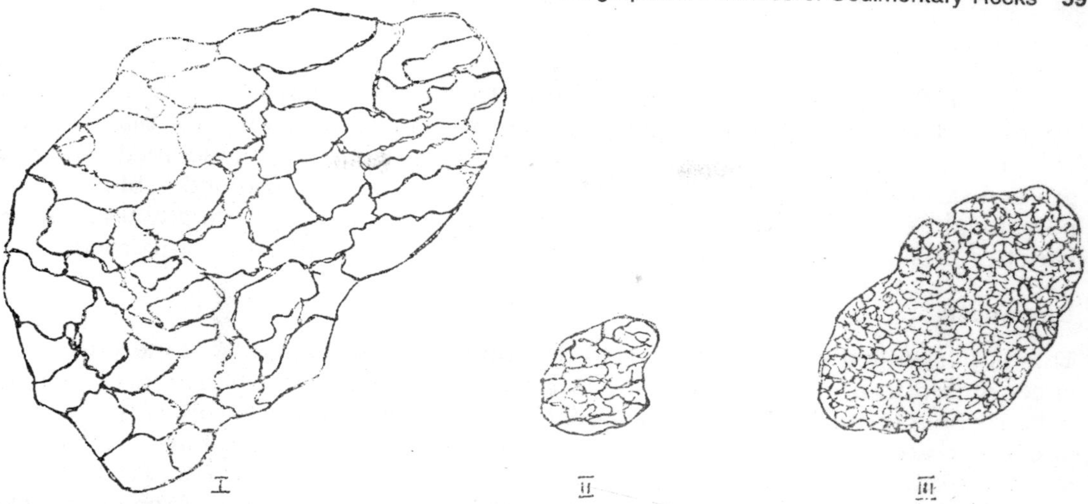

Fig. 3.14. Polycrystalline grains of quartz from quartzarenites (Camera lucida drawings ×10). *Courtesy* : G.S. Pujar (grain I); K. Bhimsen (grain II KB 139, grain III KB 115).

more than the grains of felspars. If the content of quartz be around 75 per cent, the rock is called as a sub-graywacke. If the content of quartz be between 75 and 95 per cent, it is called as a proto-quartzite. A sandstone containing more than 95 per cent of quartz, and about 5 per cent grains of chert, is called as an orthoquartzite or a quartz-arenite. In order to appreciate the subdivisions within the sandstones, a schematic diagram (Fig. 3.15) is drawn.

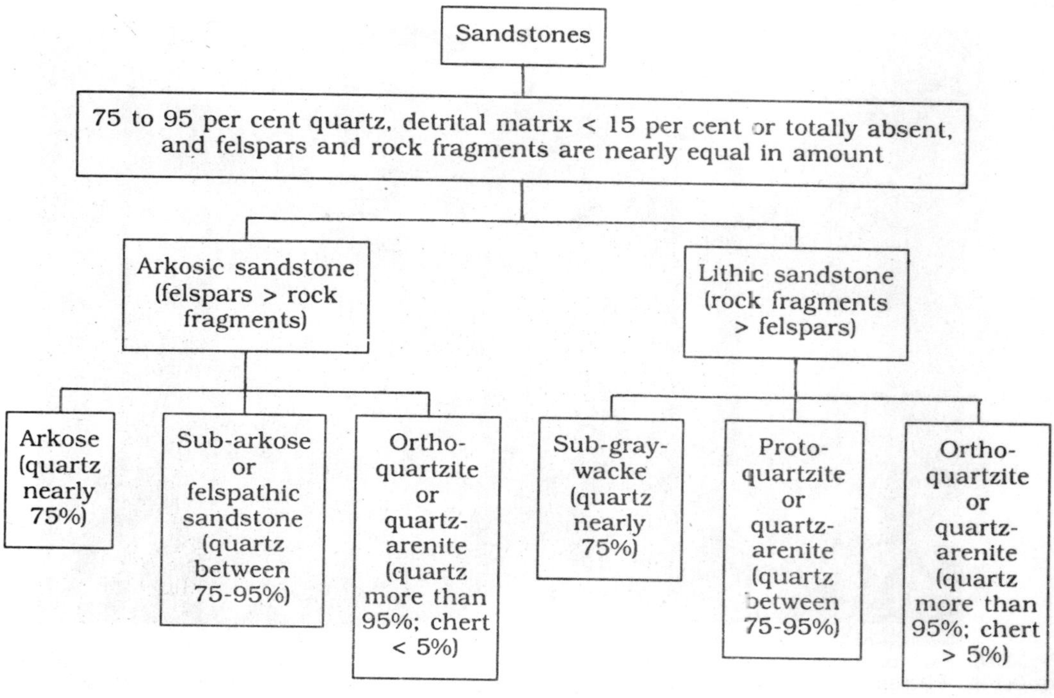

Fig. 3.15. Schematic classification of sandstones.

Whether it is a sandstone or an orthoquartzite (quartz-arenite), prefixes are added, based on the actual grain size, colour, thickness of layers or beds, external appearance and so on. Thus based on colour, varieties like cream, red, black or violet coloured sandstones are distinguished. Likewise, very coarse, coarse, medium or fine grained sandstones, thinly laminated sandstones, flagstones, unimodal, bimodal, and polymodal sandstones are distinguished. However these do not bring about any new varieties of sandstones as far as the mineralogical composition is concerned.

Graywacke

This is basically a sandstone wherein detrital matrix is considerably more than that encountered in a normal sandstone. A graywacke possesses a typical greenish gray colour which is not displayed by a sandstone. The most salient feature is the presence of considerable quantity of rock fragments which are angular in shape. The content of quartz grains is less than 75 per cent. These rocks are generally cold to "touch" in the field. Pettijohn (1957, p. 303) has noted the typical features of the graywackes as "Most striking features are therefore the dark colour of the fresh rock, the preponderance of the matrix in many specimens, the extreme angularity of the quartz and felspars, and the chaotic and random distribution of the components. The term microbreccia is most appropriate."

Graywackes are further sub-divided into felspathic, lithic and sub-graywacke varieties. In a felspathic graywacke, the content of felspar is more than that of rock fragments. When the content of rock fragments is more than that of the felspars, it is designated as a lithic graywacke. In a sub-graywacke, the detrital matrix is either absent or is very scanty (Table 3.8).

Fig. 3.16. Field photo of excellently developed gritty horizon exposed 800 metres S 25° E of Hadalge village, Ajra, Kolhapur district, Maharashtra. Note the larger dimensions of the grains and their dissemination throughout the exposure. A grit has a grain size in between that of a conglomerate and a sandstone. It is however more close to a conglomerate in its external appearance. *Courtesy :* Dr. K. Bhimsen.

Grit

This is one more variety within the group of arenites. Fragments which are angular and have a size between 2 and 1/16 mm are classified as grits. Such, grains when cemented in a matrix give rise to a "grit or a gritty sandstone". It is to be noted that the angularity of the grains of a grit, is not the outcome of any tectonic activity like faulting, dislocation, collapse, impact and so on. Bhimsen (1989) describes a gritty horizon (Fig. 3.16) in the vicinity of Nesri village, Kolhapur district, Maharashtra.

(c) Argillaceous group

Rocks composed of grains having sizes between 1/16 and below, belong to this group. Those having grains between 1/16 and 1/256 mm are called as siltstones, while those below 1/256 mm are generally called as shales, argillites and so on. Rocks of this group are rich in alumina content (Table 3.6). The relation between the several rocks belonging to the argillaceous group is shown in Fig. 3.17.

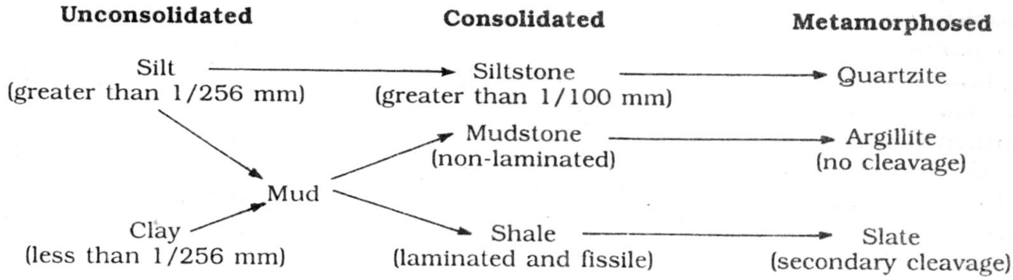

Fig. 3.17. Classification of argillaceous group (after Pettijohn, 1984, Fig. 8.1, p. 262).

Unlike the arenaceous group, this group is not divisible into sub-groups because of the fine grained nature of the components. Clay is made up of finest grade sediments (Table 2.3). Therefore minerals other than quartz, felspars (if not decomposed), a few flakes of mica, are not determinable, even in thin sections. In order to give a general idea about the quality and the quantity of minerals encountered in shales, data given by Pettijohn (1984, Table 8.3, p. 269), is set out in Table 3.9.

Table 3.9. Average mineral composition of shales (after Pettijohn, 1984, Table 8.3, p. 269)

Constituents	Clarke (1924)	Leith and Mead (1915)	Yaalon (1962a)	Shaw and Weaver (1965)
Quartz	22.3	32.0	20.0	36.8
Feldspars	30.0	18.0	8.0	4.5
Clay minerals	25.0	34.0	50.0	66.9
Iron oxides	5.6	5.0	3.0	< 0.5
Carbonates	5.7	8.0	7.0	3.6
Other minerals	11.4	1.0	3.0	< 2.0
Organic matter	—	1.0	—	1.0

It is apparent from Table 3.9 that quartz, felspars and clay minerals practically make up the entire composition of the shales. Further, the grains of felspars are often found to be decomposed to clay minerals. Clay minerals (like kaolinite, illite etc.) are identifiable by the X-ray analysis. In thin sections therefore only quartz and flakes of mica can be identified. Therefore it may be said that the clay minerals very much predominate over quartz, in shales.

Therefore the argillaceous rocks are mostly classified on the basis of colour, fissility, and not on the minerals composing them. Argillite is a hardened shale which lacks cleavage, as described by Grout (1932, p. 365). Shale is characterised by the presence of fissility. Thus varieties like the laminated shales, the varved shales and the fissile shales are distinguished. Concretions possessing fanciful shapes are not uncommon in the shales and the argillites. These rocks also display considerable variation of colours. White, brown, reddish brown, yellowish brown coloured varieties are noticeable. Variegated coloured shales and argillites are excellently exposed at and around Dharwad, Karnataka state.

(d) Calcareous group

This group is characterised by the presence of carbonate minerals of calcium and magnesium, in the main, those of iron forming a very small amount in these rocks. Rocks made up of $CaCO_3$ predominate over those of $CaMg(CO_3)_2$, the rock names being limestone and dolomite, respectively. Minerals composed of $CaCO_3$ are calcite and aragonite, $CaMg(CO_3)_2$ builds up dolomite, while $FeCO_3$ forms siderite. Other constituents which form a very small proportion of these rocks are quartz, clay minerals, felspars, glauconite, collophane and pyrite. The distinction between a limestone and a dolomite is made on the basis of the content of $CaMg(CO_3)_2$ in the limestone, and that of $CaCO_3$ in the dolomite. This has been shown in Table 3.10.

Table 3.10. Classification of carbonate rocks

Content of calcite ($CaCO_3$) →			← Content of dolomite $CaMg(CO_3)_2$		
100%	95%	90%	50%	10%	0%
	Limestone	Magnesian limestone	Dolomitic limestone	Calcitic dolomite	Dolomite

Limestones

Limestones proper contain about 5 per cent of $MgCO_3$ in them. Those with 5 to 10 per cent of $MgCO_3$ in them, are called as magnesian limestones, while those containing between 10 and 50 per cent of $MgCO_3$, are classified as dolomitic limestones. The grain size of the calcareous sediments is generally very fine, like that possessed by the clays. However there are varieties with coarser grain sizes too. Calcareous rocks having a grain size between 2 and 1/16 mm are called as calcarenites, while those having grain size greater than 2 mm are termed as calcirudites. Unlike the arenaceous and the argillaceous rocks, organisms give rise to limestones. Depending upon the kind of the organism, different varieties are recognised like the foraminiferal limestone, the coral limestone, the stromatolytic limestone and so on.

The limestones display a variety of textures. Mechanically formed limestones possess current sorting and stratification, cross bedding, graded bedding. Biochemically formed limestones display algal structures, stromatolytic structures, algal banding. Oolitic and pisolitic structures are produced in the chemically deposited limestones. The limestones abound in replacement structures like the stylolitic seams, the relict structures and so on. The sub-divisions made within the limestones on the basis of grain size, texture and the presence of fossils, are presented in Table 3.11.

Table 3.11. Classification of limestones based on grain size, textures and fossils

Limestones			
Grain size in mm		Kind of fossil	Kind of texture
Greater than 2	Between 2 and 1/16		
Calcirudite	Calcarenite	Fossiliferous limestones, foraminiferal, coral, stromatolytic limestones	Pisolitic, oolitic, stylolitic limestones

The limestones are also classified on the basis of their origin. Some are fragmental or detrital, and these are mechanically transported and later deposited. Others are chemical or biochemical precipitates. In Fig. 3.18, the classification of the limestones based on origin, is presented.

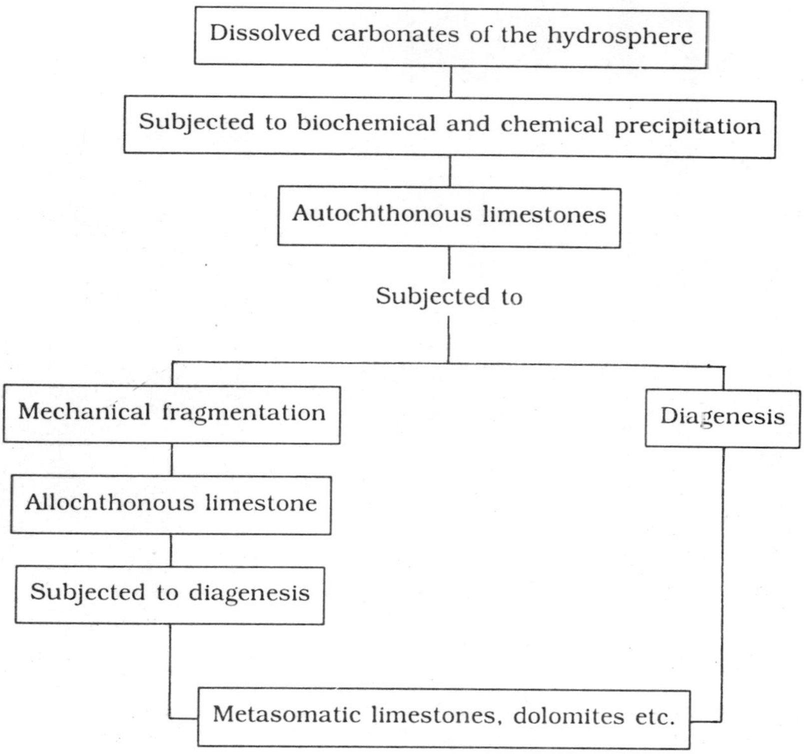

Fig. 3.18. Schematic classification of limestones (after Pettijohn, 1957, Fig. 90, p. 382).

Fig. 3.19. Field photo showing a typical laterite/bauxite profile. Note the typical reddish brown colour of the laterite (top portion of the outcrop), and the light gray colour of the bauxite (towards the bottom part of the outcrop). The quarry is one kilometre east of Vadavli village, Shrivardhan, Kolaba district, Maharashtra state. *Courtesy :* Dr. A.H. Kouhsari.

Fig. 3.20. Field photo of a quarry in bauxite, the locality being Kuravde village, Shrivardhan, Kolaba district, Maharashtra state. Note the colour difference between a bauxite (it is grayish white in colour), and a laterite (which is reddish brown in colour, as seen in Fig. 3.19). *Courtesy :* Dr. A.H. Kouhsari.

It is seen from Fig. 3.18 that three types of limestones namely allochthonous, autochthonous and metasomatic varieties are distinguishable. The autochthonous limestones are further regarded as the "primary rocks", from which both the allochthonous and the metasomatic types are produced.

Dolomites (dolostones)

A dolomite is a magnesium carbonate rock containing 10 per cent of calcite in it. When the content of calcite is between 10 and 50 per cent, the rock is named as a calcitic dolomite (Table 3.10). The name dolomite is also applied to a mineral. Therefore to avoid confusion, the name "dolostone" is used for the rock, while the name dolomite is used for the mineral. Dolomite displays textures similar to those encountered in the limestones, but the former rock (dolomite) is more even grained. The dolomites are produced by the replacement of $CaCO_3$ by $MgCO_3$, a process called as "dolomitisation". As a result of this process, the original texture of the limestones gets steadily obliterated. This gives rise to relict or the ghost texture, i.e., left over portions of the original parental texture of the limestone in the newly formed dolomitic rock. Weathering of a dolomite produces a buff coloured material, and this field feature may be utilised to distinguish a dolomite from a limestone. Ferrous carbonate which is held in solid solution in a dolomite, on oxidation, it gives rise to the buff coloured product which is so characteristic of the dolomitic rocks. Calcite does not hold any ferrous carbonate in solid solution within it, and therefore it does not give rise to buff coloured product on weathering during oxidation of the limestones.

(e) Oxides and hydroxides group

This is a very small group belonging to the sedimentary rocks. It differs from the other groups in that the rocks of this group are not formed in the sedimentary basin. The insoluble products of weathering are left behind at the site of weathering. The dominant product is made up of oxides and hydroxides of iron and manganese. Some grains of quartz are also left behind. Such rocks are called as laterites (Fig. 3.4). These display oolitic and pisolitic structures. Laterites are generally reddish brown in colour (Fig. 3.19) due to the dominance of iron oxide in them. In some instances, manganese predominates and gives rise to manganiferous laterite. When alumina (Al_2O_3) dominates over manganese, it produces an aluminous laterite, which is technically called as a bauxite (Fig. 3.20). There is yet another product of weathering which does not reach the basin of sedimentation. It is the soil. As it consists of clay minerals, it is to be regarded as a rock, and to belong to the group of "oxides and hydroxides".

SUMMARY

Below noted features are worthy of enumeration :

1. These are almost monomineralic rocks, one or two minerals composing the entire rock.
2. Frequently occurring minerals are quartz, felspars, clay minerals, calcite, dolomite and muscovite.
3. Sandstones, shales, limestones, dolomites (dolostones), conglomerates, grits and graywackes occur in the decreasing order of abundance.

4. Based on absolute grain size, rocks recognised are rudites, arenites, and argillites. These are family names.
5. Based on mineral composition, rocks recognised are sandstones (quartz), shales (clay minerals), limestones (calcite), dolostones (dolomite and calcite).
6. Based on textures/structures rocks recognised are grit, breccia, laterite, flagstone and conglomerate.
7. Numerous heavy minerals are encountered, though these form about one per cent of the entire rock.
8. Fossils are found exclusively in these rocks.
9. Mineral grains are held together in a matrix.

CHAPTER 4

Processes and Environments of Sedimentation

INTRODUCTION

The several geological agents like wind, running water, glaciers and so on, bring about transportation of the products of physical and chemical disintegration and decomposition of the crustal rocks, towards the seas, lakes and the oceans. It is said towards the seas and the oceans, because some of the products are left behind at the site of weathering, like a laterite, a bauxite and so on, or give rise to soils. However ultimately all the products of disintegration and decomposition reach the seas and the oceans, because these places mark the lowest levels on the surface of the earth. Technically the products of disintegration and decomposition are called as the "sediments". These comprise of soluble and insoluble minerals to which a mention has been made in an earlier chapter. Sedimentation, meaning deposition of the products of weathering into a basin (like sea, lake or ocean), will depend upon the nature of transportation, i.e., whether in suspension or in solution. The kind of sedimentation will depend upon what is being brought into the basin. Running water (streams, rivers) and wind are the main agents of transportation, and hence these are of deposition too. Glaciers especially transport very large sized blocks of rocks. During the disintegration and decomposition of the crustal rocks, firmly held minerals are loosened, and a combination of newly formed minerals (like clay) and released ones (like quartz, heavy minerals) and smaller sized rock pieces, is produced. But during this process, the compact and the hard crustal rocks get broken down to smaller sized granular material. Such a heterogeneous mass (in the sense of varying sizes of the individual constituents) will be on its way to the basin. Depending upon the length of transport, further reduction in the sizes of the constituents of the sediments is brought about by the wear and tear (called attrition). However some large pieces of rocks escape the wear and tear, because of their hardness and resistance. These however get rounded to sub-sounded in shape, and reach the basin as pebbles, cobbles, boulders and so on.

It will be appreciated that the kind of sediments (whether coarse, medium or fine grained material brought in suspension, or in solution) depends upon the nature and the distance of transportation undergone by them (sediments). The transportation ceases as soon as the sediments reach the basin, and the actual process of sedimentation commences.

AGENTS OF TRANSPORTATION AND THE KIND OF SEDIMENTATION

The products of disintegration and decomposition of the crustal rocks get differentiated or fractionated into different categories mainly on two properties namely, (a) density, and (b) solubility (in water) of the constituents composing the sediments. At the site of weathering, the lighter, the denser, the soluble and the insoluble materials all of them occur together. It is the agents that bring about the separation between them. The streams, the rivers and the wind, carry the lighter and the smaller sized materials more easily than the heavier and the bigger sized ones. The glaciers are capable of transporting extremely large sized rock masses alongside with finer ones too. But this agent is unable to bring about a separation between the lighter and the denser materials. The glaciers do not carry the materials in suspension as is done by the rivers and the wind. Thus from the nature of the sediments reaching the basin, the agent of transportation could be inferred with a fair degree of accuracy. A sort of a ready reckoner in this regard is presented in Table 4.1.

Table 4.1. Features of sediments transported by different agents

Agent of transportation	Features of sediments transported
Wind	(a) Grains are very fine and are considerably well rounded.
	(b) A better separation is effected between grain sizes. Well assorted grains are met with.
	(c) Rudaceous material is not encountered.
	(d) Only resistant grains like those of quartz are met with.
	(e) Calcareous sediments are not met with.
Streams and rivers	(a) All four types of sediments namely rudaceous, arenaceous, argillaceous and calcareous are met with.
	(b) A better separation is expected because the sediments are carried in two different states—in suspension, and in solution. Also the distance of transport is more as compared to that associated with the glaciers. Lighter, denser and the soluble materials, hence, get better separated.
Glaciers	(a) Extremely unassorted sediments.
	(b) Very large sized rock masses, pieces, and blocks of rocks are transported.
	(c) Grains are mostly sub-angular ones.
	(d) Undecomposed primary minerals like felspars are possible in the sediments.
	(e) Only one face of the rock-pieces may get striated, polished or smoothened.
	(f) Very fine varve structures are met with.

It is apparent from Table 4.1 that the kind (quality) of sedimentation (sediments) depends considerably upon the kind of the agent of transportation. Quantity of only one kind of sediment (whether rudaceous, arenaceous, calcareous or argillaceous) will depend to a considerable extent on the provenance (meaning the rock undergoing weathering), but the agent of transportation also contributes to it.

ENVIRONMENTS OF SEDIMENTATION

The final and the actual site of deposition (sedimentation) of the sediments, is the basin. The water in the basin may be deep or shallow, or the depth might vary during the time span of sedimentation. The floor of the basin may be flat, or it could be undulating one. The water in the basin may be quiet or there could be currents generated of varying intensities. The water may be fresh or saline, it may be charged with CO_2, H_2S, O_2 and so on. This is spoken of as the oxidising and the reducing environments. All these features together build up the environment of sedimentation. Chemical deposition (precipitation) of carbonates held in solution, needs a favourable chemical environment.

REALMS OF SEDIMENTATION

The most fundamental factor controlling the quality of the sediments is the tectonics. Diastrophism produces elevations (mountains, hills etc.) and depressions (lakes, seas and oceans) on the surface of the earth. As a result of this, the geological agents start their activities of erosion of the elevated parts of the crust of the earth, and of deposition of the products of weathering into the depressions, in order to bring about a graded surface for the earth. Without erosion, there cannot be any sedimentation. Diastrophism is hence the cause of sedimentation. The rate of erosion and the consequent sedimentation, are directly controlled by the nature of tectonism. Tectonics controls the relief (undulating surface) of a region, while the relief controls the rate of erosion. Greater the relief, faster will be the rate of erosion. When the erosion is fast, products (sediments) of erosion will be large in quantity. The rate of sedimentation depends upon the rate of supply of the products of erosion (sediments), and the rate of subsidence of the basin in which the sediments are being laid. It is important to note that the rate of supply of the sediments and the rate of subsidence of the basin, are entirely controlled by the tectonics.

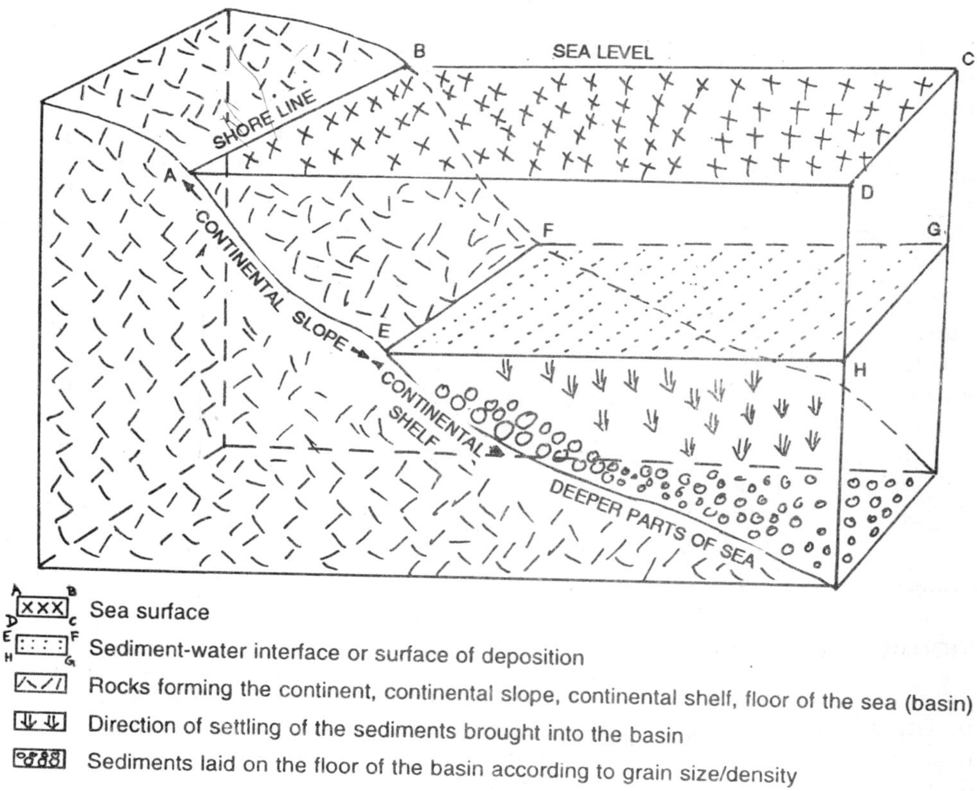

Fig. 4.1. Two surfaces (sea level and sediment-water interface) controlling nature of deposition and texture/ structure acquired by the sediments. These later become sedimentary rocks.

It is necessary to make a mention to two surfaces having a bearing on the mode of sedimentation. *One is the surface of deposition, and the other is the sea level* (Fig. 4.1). Depending upon the situation, sediments are deposited either above or below the "wave base" (Fig. 4.1). Wave base means the lower limit up to which the wave motion is effective. Up to the wave base, the water of the basin is found to be turbulent, while below the wave base, the water is found to be quiet. Therefore the structures developed in the sediments deposited in these regions will be different. Even the type of sediments laid, and their thicknesses could be different. The diastrophism could be periodic or episodic. Therefore the establishment of equilibrium between the "erosion and weathering or supply of sediments", on one hand, and the "rate of subsidence of the basin" on the other, might get upset. However the re-establishment of equilibrium always proceeds in an orderly manner and gives rise to an orderly sequence of sedimentary facies (rudaceous, arenaceous etc.).

Tectonics plays yet another role. It influences the maturity of the sediments. But it has been observed that the relief rather than the climate controls the maturity (content of grains of felspars) of the clastic sediments. A high relief promotes the removal of the incompletely decomposed primary minerals. Such sediments therefore carry undecomposed laths of felspars, rendering the sediments immature. On the other hand, a low relief favours complete decomposition of the felspars. Such sediments therefore do not contain undecomposed felspar laths in them, and are regarded to be mature ones.

The different realms of deposition are dependent on the relation between the "surface of deposition" and the "sea level". Generally the former surface is at a level lower than the sea level (Fig. 4.1), but it might also stand at a higher level. The surface of deposition is primarily dependent on the total quantity of sediments reaching the basin, as well as on the rate of subsidence of the basin. Strongly negative areas (these form the geosynclines) are indurated and covered with deep marine waters. Turbidity underflows give rise to rhythmic and graded bedding. Such an environment produces "flysch facies" of sediments which later give rise to flysch facies of sedimentary rocks.

Accelerated rate of sedimentation or a decreased rate of subsidence of the basin, results in the depositional surface being raised to or above the sea level. Under such an environment, "mollase type" of sedimentation takes place. When the rate of subsidence of the basin is considerably high, but the rate of supply of the sediments is poor, it produces the "euxinic type" of environment. Under this environment, only chemical deposition (precipitation) takes place. When tectonism is absent, it produces stable surface of deposition. Under such an environment, carbonate deposits like the limestones, the dolomites etc. are formed.

The diastrophism gives rise to two main realms of deposition, namely (i) the craton, and (ii) the geosyncline. The cratonic basin is also called as the "shelf region" of the ocean. By and large, "oxidising environment" is prevalent in a craton, and orthoquartzites, carbonatites, and evaporites are formed in such basins. In a geosyncline, the environment is "reducing" one, and euxinic, flysch and molasse type of sedimentation is encountered. These features have been presented in Fig. 4.2. Very broadly it may be said that :

(i) Lithology, relief and the climate of the source rock area (provenance) determine the composition of the sediments.

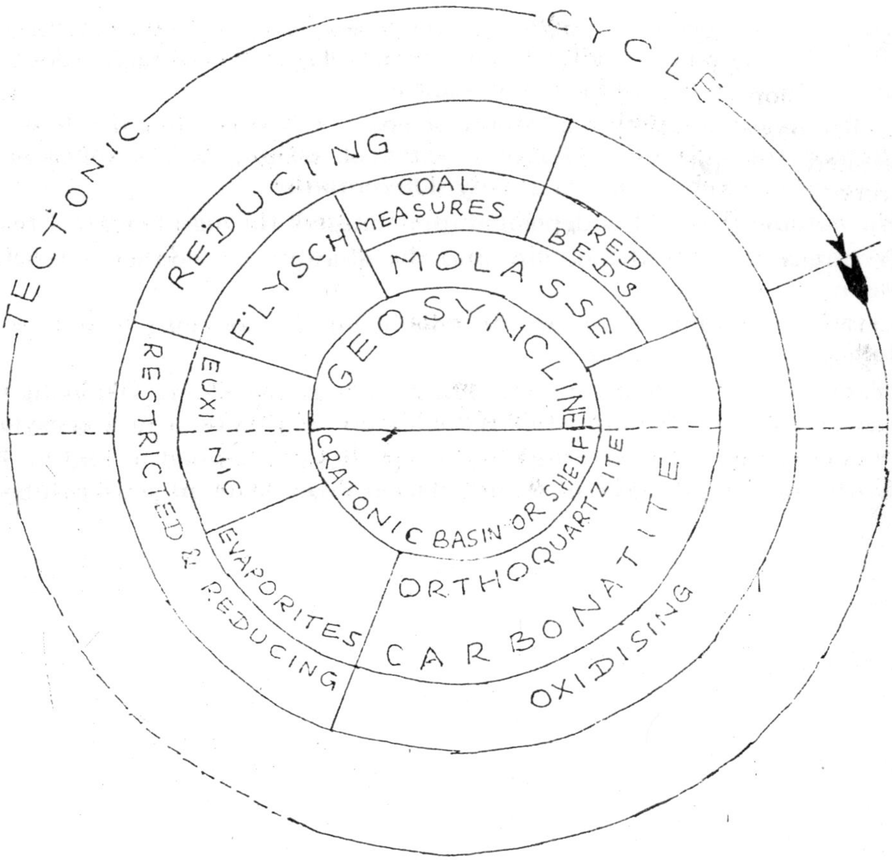

Fig. 4.2. Tectonic cycle and environments of sedimentation (after Fig. 164 of Pettijohn, Sedimentary Rocks, 2nd edition, 1957).

(ii) Environment of deposition controls the textures, the structures and the geometry of the sediment-bodies.

(iii) Tectonism exerts an overall control over the major sediments reaching the basin because it influences

(a) the relief of the source area (provenance), and

(b) negativity of the site of deposition.

SUMMARY

Below noted features are noteworthy :

1. Disintegration and decomposition (weathering) of the elevated parts of the crustal rocks, create soluble and insoluble materials. This initiates the process of sedimentation.
2. Rivers, streams, glaciers and wind transport the products of weathering to the sites of deposition (basin). The insoluble products are either left behind or are carried in suspension. The soluble ones are transported in solution by the streams and the rivers. Some sediments are carried in a colloidal state.

3. During transport, a separation of the heavier grains from the lighter ones is effected. Also, further reduction in the size of grains is brought about through the collision of grains upon one another.
4. In the basin, further separation of coarser grains from the finer ones is effected through the differential rates of settling of the different grains. Process operative is one of gravity differentiation.
5. The soluble material is deposited (precipitated) through chemical reactions.
6. Tectonics controls the quality and the quantity of sediments reaching the basin.
7. Realms of sedimentation are dependent on the positions of the "surface of deposition" and "sea level".
8. Craton and geosyncline are the two realms of deposition. Oxidising environment exists in a craton, while reducing one is prevalent in a geosyncline.
9. Orthoquartzites, carbonatites and evaporites are formed in craton. Euxinic, flysch and mollase type of sedimentation takes place in geosynclines.

CHAPTER 5

Origin of Sedimentary Rocks

INTRODUCTION

The formation of the sedimentary rocks is an abstract aspect similar to that of the formation of the igneous and the metamorphic rocks, because in the case of the sedimentary rocks also, the actual modus operandi (meaning the method of operation) cannot be demonstrated in the laboratory. Several features like the mode of occurrence, the mineralogical composition, and the textural/structural features are required to be taken together, because all of them indirectly indicate the process or the processes of origin of the rocks. It is to be remembered that in respect of the mode of occurrence, the mineralogical composition, and the textural/structural features, the sedimentary rocks differ considerably from those of the igneous and the metamorphic rocks. Therefore the mode of formation of the sedimentary rocks is bound to be different one. While describing the distinctive features of the sedimentary rocks (Chapter 1), it has been mentioned that these are derived from the pre-existing crustal rocks. Hence the first pre-requisite to be considered is the nature of the pre-existing rocks. Another important aspect is that the sedimentary rocks mark (occur) the sites of basins of the geological past. Therefore without the existence of the basin or basins, the sedimentary rocks cannot be formed. It may therefore be said that a favourable combination of pre-existing rocks, and a basin, controls in the main, the origin of the sedimentary rocks. These and the other aspects will be elaborated in describing the manner of development of these rocks. The evolution of the sedimentary rocks is presented in the form of stages, and it has been shown how the rocks are steadily and systematically developed.

STAGES IN THE FORMATION OF SEDIMENTARY ROCKS

Six stages are recognisable in the evolution of the sedimentary rocks. 1st stage is the disintegration and decomposition of the crustal rocks. 2nd stage is the transportation of the products of disintegration and decomposition by wind, river, glaciers and other agencies. 3rd stage is the formation of the basin. 4th stage is the deposition of the products of disintegration and decomposition brought about by the several agencies into the basin. 5th stage is the separation of the sediments brought into the basin, into the rudaceous, the arenaceous, the calcareous and the argillaceous fractions. 6th stage is the compaction of the various sediments into hard rocks. Final stage is the upliftment of the basin in which the sediments were laid earlier. Each stage is elaborated in the following paragraphs.

(a) Disintegration and decomposition of crustal rocks

This constitutes the *first stage* in the origin of the sedimentary rocks. Unless the crustal rocks get elevated to considerable heights above the sea level, action of the agents of weathering cannot be initiated. Therefore some parts of the crust of the earth undergo upliftment, and such regions form the provenance for the sediments. Such elevated crustal parts are technically called as the "geanticlines", which in all probability form the adjoining counter part of the "geosynclines" (which form the basins). Whatever might be the cause of the upliftment of certain parts of the crust of the earth (because many hypothesis have been proposed in this regard, like the intrusion of an igneous body, the plate tectonics, and so on), suffice it to say that such elevated parts forming the hills, the mountains, valleys and so on, the rocks constituting them, undergo disintegration and decomposition, and give rise to the sediments. It is of interest to draw a parallel between the role of magma in the origin of the igneous rocks, and that of the provenance in the origin of the sedimentary rocks. Without a magma, the igneous rocks cannot be formed, though difference of opinion exists, regarding the generation of the magma itself, amongst the petrologists. Further, depending upon the nature of the magma, the resultant igneous rocks also vary. Likewise it may be said that depending upon the composition of the elevated crustal rocks (this forms the provenance) undergoing weathering, the products derived therefrom (meaning the sediments) will logically vary in composition. Of course, besides the provenance, the intensity of weathering and its nature, and the distance of transportation undergone by the sediments, also govern the quality and the quantity of the transported material reaching the site of deposition (the basin). It is therefore said that much depends upon the composition of the rocks undergoing weathering, and the distance of transportation undergone by the sediments, in building up the composition of the resultant sedimentary rocks.

(b) Nature of transportation

This is the *second stage* in the evolution of the sedimentary rocks. The main role played by it is in bringing about a separation of the products of weathering. The insoluble ones like the iron oxide, the manganese oxide, and the alumina (Al_2O_3) are either left behind, or are transported in suspension to relatively shorter distances away from the provenance. It is thus quite obvious that the parental basic rocks like the gabbros, the basalts, the pyroxenites etc., and not the acidic and the intermediate igneous rocks (like the granites, granodiorites, syenites, pegmatites etc.) could be expected to produce iron rich laterite, because such rocks are relatively richer in the contents of iron. Likewise the acidic and the intermediate igneous rocks, being richer in the contents of felspars and quartz, these are more prone to give rise to the aluminous laterites (bauxites) and not to the iron laterites.

The other insoluble material carried in suspension is silica (grains of quartz). During the transport, these grains get rounded and reduced in size. Greater the distance of transport, finer and more rounded will be the resultant grains. It is therefore reasonable to expect the fine grained sedimentary rocks like the sandstones and the shales to be formed in basins that are situated far away from the source rocks undergoing weathering. It is important to note that the grains transported by the wind are extremely fine and are very well rounded in shape.

Actually the transporting agents reduce the size of the grains, render them rounded to sub-rounded in shape, and give rise to sediments called as rudites, arenites and argillites. However in the basin, further fractionation (meaning separation) of sediments, according to the grain size takes place, finally giving rise to the rocks called as the conglomerates, the sandstones, the shales and so on. Much depends upon which agent had transported the sediments. Wind and the rivers bring about an almost complete fractionation (separation) of the sediments into rudaceous, arenaceous and the argillaceous categories. The glaciers also are active agents of transportation, but they effect least separation of the sediments according to the grain size. It, therefore, produces a heterogeneous mixture of giant sized blocks of rocks, and very fine grained sediments, at one and the same place. The glaciers may be described as the "bulk carriers" of the products of weathering.

It is thus clear that the nature of the transporting agent and the distance of transportation of the sediments away from the provenance, have far reaching effects upon the composition of the resultant sedimentary rocks.

(c) Formation of a basin

Whereas the provenance is formed through the upliftment of a part of the crust of the earth, a "basin" on the other hand needs a sinking or a subsidence of a part of the crust of the earth. Whether these "two acts (episodes)" are taking place simultaneously, is a debatable matter. How much far apart these two sites (provenance, and the basin) are situated, is again a debatable matter. It is generally held that strong negative gravity anomalies result in the subsidence of a part of the crust of the earth. The downsinking of the convection currents (Holmes, 1987) also can initiate a basin. Therefore, instead of enlisting the various hypotheses in respect of the formation of a basin, let us turn our attention to the role played by it (basin) in the formation of the sedimentary rocks. Depending upon the depth of water in the basin, it is logical to expect the thickness of the sediments laid, to vary. Deeper the basin, thicker the sediments laid, shallower the basin, thinner will be the sediments laid. Shallow depth of water also favours development of structures like the current bedding, the ripple marks, the rain prints and so on. Deeper and quiet water is congenial to the development of graded bedding, formation of shales and limestones. If the basin be marshy one, the prevailing reducing environment favours the development of flysch and molasse type of deposits. If the depth of water be fluctuating one, then it will give rise to thick (more depth) and thin (shallow depth) layers of sediments, which in turn result in thick and thin beds of rocks. In this way the basin controls the thickness of the sediments laid, as well as the environment of sedimentation (reducing, oxidising etc.). Formation of a basin marks the *third stage* in the evolution of the sedimentary rocks.

(d) Deposition of the products of disintegration and decomposition

This forms the *fourth stage* in the evolution of the sedimentary rocks. There is a definite order in which the products of disintegration and decomposition (these give rise to the sediments) are laid in the basin. As soon as the river enters the basin, its velocity decreases, and the sediments carried by it in suspension, cease to travel any further distance. The first to settle down on the floor of the basin, are the large sized materials namely, the boulders, the cobbles, the pebbles and the like. These settle down on the floor of the basin first, because these are relatively heavier than the "arenaceous and the argillaceous grade materials". A layer or a bed of boulders, cobbles or the pebbles will be found at the "shore" or the "marginal" portion of the sea (basin). The arenaceous material is likely to be carried over, to more distance into the sea (basin) by the action of the sea waves, and possibly by the slackened onward movement of the river water. During this stage, the "arenaceous grade materials" settle down, and form a layer or a bed over the rudaceous material laid down earlier. In this way, the conglomerates and the sandstones (beds) come into existence.

The "argillaceous grade material" being very fine grained, it takes more time to settle down. This can happen off the coast, and in the deeper parts of the sea (basin) where the water will be considerably quiet, and not affected by the effect of the sea waves. Analogy can be given of the inland lakes. The water of such lakes is observed to be muddy during the rainy seasons. But during the drier periods, the same water is found to become clear, because the small clay particles settle down to the bottom of the lake, producing a layer of mud. This process is technically called as the silting of the lakes. Similarly, the river water is muddy during the rainy season, and also when it is in spate. But during the other seasons, the river water is again found to become clear, because the clay particles carried by the river settle down on the floor of the river. In the basin of sedimentation, a similar process can be expected to be operative, especially for the finer grained clay particles. In the basin, the resultant layer gives rise to a shale or a mudstone on the floor of the basin.

The "rudaceous, the arenaceous and the argillaceous grade materials" are carried in suspension, and these reach the basin in that state. The process of deposition of such a material is through gravity; the denser material settling first (pebbles, cobbles, boulders of conglomerates), the lighter material (arenaceous material) settles next, and finally the lightest material (clay particles of shales). The separation of these fragments (rudaceous, arenaceous and argillaceous) is through the differing settling rates—denser material settling faster, and the lighter one, settling very, very slowly. This behaviour is described as the process of "sedimentary differentiation".

The calcareous material ($CaCO_3$, $MgCO_3$, $FeCO_3$) reaches the basin in a dissolved state and its deposition requires chemical activity. It requires an environment totally different from that required for the deposition of the rudaceous, the arenaceous and the argillaceous material. The water is required to be quiet, and this is realised at greater depths of water (meaning deeper parts of the basin). Chemical reactions are also necessary, and only then the carbonates of Ca, Mg and Fe could be formed.

Some of the sediments are carried in a colloidal state. Such a sediment gets coagulated in sea water due to the presence of salts, which act as electrolytes. pH of the basinal water also plays an important role in the process of precipitation. Thus manganese and iron get separated during deposition.

As the deposition takes place in stages, it gives rise to layers or beds of sediments which differ in grain size, as well as in mineral composition. It also takes place in a definite order. There are certain features which remain ambiguous. Do the sediments reach the basin in bulk, and only then the sedimentation takes place? Or do the rivers dump the sediments during different periods, and in installments? Does the sedimentary differentiation operate during each period of dumping of the sediments? Do the rivers pour in the sediments from all sides of the basin, or only from one or two sides? Whether a few rivers carry out the act of bringing in of the sediments to the basin? There is no clarity regarding these several aspects amongst the sedimentologists. Whatever the actual fact might be, the process of sedimentation is one of gravity settling of the insoluble sediments, and the chemical precipitation of the soluble sediments.

(e) Separation of the products (sediments) brought into the basin

This is the *fifth stage* in the evolution of the different varieties of sedimentary rocks. In actuality, the separation or the differentiation commences right from and at the site of rocks undergoing weathering. The insoluble material gets separated from the soluble ones. The insoluble material comprising of the oxides of iron, aluminium and manganese later gives rise to the rocks called as the laterites, to which a mention has already been made. This material of the laterites does not reach the basin of sedimentation, and hence it is not found along with the rocks formed in the basin, like the sandstones, the conglomerates, the shales and so on.

Grains of quartz and fine particles of clay, flakes of mica are also insoluble, but these are transported by the rivers, the streams, and the wind, to the basin, in the state of suspension. The soluble material comprises of carbonates of Ca, Mg, Fe and these also reach the basin. Further separation takes place in the basin, which is based on the differences in the rates of settling of the sediments. As a combined effect of the transportation by the rivers and the wind, and the differential rates of settling, the sediments brought into the basin get more or less perfectly separated into the rudaceous, the arenaceous, the argillaceous, and the calcareous fractions, and these are deposited as separate layers, one over another, in an orderly manner. Rivers and the wind initiate the separation of the sediments, and it is continued further, and is carried out to completion, in the basin. A longer distance of transport coupled with a prolonged period of sedimentation in quiet water, ensures a better separation of the sediments.

(f) Compaction of sediments to form rocks

In the *sixth stage*, the loose ingredients (grains) of the different layers laid in the basin, are compacted to form hard rocks. The processes called into play are technically called as "diagenesis and lithification". The term diagenesis stands for the reactions taking place within a sediment, between one mineral and the other, or between one or several minerals and the interstitial or supernatant fluids (Pettijohn, 1957, p. 648). The coarse, the medium and the fine grains of the sediments are merely lying close to each other, and these are further soaked in the water of the basin. Mention has been made to the "surface of deposition" existing in the basin. It means that, above such a level, there is plenty of water which is devoid of the sediments. But we are familiar with the sedimentary rocks that are hard and compact, which in fact are required to be broken with a hammer, in order to obtain a sample for study. In the following paragraphs, the transformation of the loosely held grains of the different sediments, into hard and compact rocks has been described.

The constituent grains of the sediments are required to be held together by some binding material. It appears that the finer grains of the sediments act as a cementing material to firmly bind the several bigger grains together. The finest materials are the clay particles and generally these form the cementing material. The finer siliceous (quartz) grains also act as a binding material. At times, the ferruginous and the calcareous material also forms such a material. Thus the compaction and the hardening of the sediments is, by and large, brought about by the finer grained material, which is also brought into the basin by the rivers and the streams and the wind.

It is also necessary to expel or squeeze the water out from the sediments, because, even if the cementing material is trying to hold the grains of the sediments together through the cohesive forces, but the water present in between the grains, renders the sediments loose and uncompact. Usually large thicknesses of sediments are laid one over another, as layers or beds. In a geosyncline for example, 4000 to 5000 metre thick sediments appear to have been laid, judging from the sedimentary rocks disclosing such thicknesses in the Himalayan mountains, the Alps and the other sedimentary basins. The mere weight of such a large pile of sediments, must be pressing out the water held in the spaces in between the grains of the sediments.

Even though the sedimentary rocks are formed practically at the surface of the earth, as compared to the depths at which either the igneous or the metamorphic rocks are formed, it appears that the sediments laid in the basin do undergo a very mild and a low degree of recrystallisation (metamorphism), because these (sediments) are taken down to slightly deeper parts of the crust of the earth, where slightly elevated temperatures and pressure could be logically expected to exist. This environment is especially feasible in the geosynclines, because it is argued by the geologists that such basins sink in order to accommodate large quantity of sediments brought into them. Due to this, there will be rise in both temperature and pressure, which brings about a low grade metamorphism (recrystallisation) of the sediments. This process also brings about increased coherence between the loosely held grains of the sediments.

Microscopic examination of the sedimentary rocks reveals growth around some grains of quartz and felspars (Figs. 3.2 and 3.3). This is called as the process of "authigenic overgrowth". Such an overgrowth takes place after the grains are cemented together, producing a layer or a bed. The role of the overgrowth is to further strengthen and strongly hold together the several grains of the sedimentary beds or layers. Likewise, the replacement of grains also takes place which introduces new minerals like tourmaline and others, into the sedimentary rocks. This process is operative through the chemical reactions and through the process of gravity settling of the sediments.

GENESIS OF THE DIFFERENT SEDIMENTARY ROCKS

From the foregoing account, it is clearly appreciable that the composition of the sedimentary rocks originates through the joint action of the different sediments brought into the basin by the rivers, wind and glaciers, and being further separated into the rudaceous, the arenaceous, and the argillaceous fractions, through the differential rates of settling of the sediments. The calcareous fraction is precipitated through the chemical action. The rudaceous material gives rise to the conglomerates. If the sizes of the pebbles be too large, it gives rise to the rock called as the "boulder bed". Conglomerate is the first rock to be formed in the basin, and it marks the border and the shallowest part of the basin. The arenaceous sediments give rise to the sandstones. Depending upon the actual sizes of the grains, varieties like the coarse, the medium and the fine-grained sandstones are formed. Sandstones mark the second phase of sedimentation, and these rocks are formed in the relatively interior (deeper) parts of the basin, as compared to that of the conglomerates.

The argillaceous sediments give rise to the clay rocks like the shales, the argillites, the mudstones and so on. These are formed in the further interior (deeper) parts of the basin where the water is deep and also quiet. The calcareous sediments which are carried in solution to the basin undergo chemical reactions, and the carbonates of Ca, Mg and Fe are precipitated. This results in the formation of the limestones, the dolomites (dolostones) and so on.

DEVELOPMENT OF DIFFERENT STRUCTURES

The textures and the structures of the sedimentary rocks are controlled by the intensity and the nature of the transportation undergone by the sediments. Longer the transportation, finer will be the grain size. If the crustal rocks (provenance) be already fine grained ones, then even a shorter distance of transportation of the sediments can give rise to the finer grain size. In addition to this, the depth of water in the basin also affects the resultant textures and the structures of the sediments laid in the basin. Quiet water permits the separation of the coarse from the fine grained sediments. Thus the coarser ones settle down first followed by the finer ones, in a systematic manner. If at this stage, a fresh quota of sediments reaches the basin, once again the coarser sediments settle down first, followed by the finer ones. In this manner, the *graded bedding* is produced in the sandstones.

If the depth of water be very shallow, the waves produced in the water of the basin produce *current and the oscillation ripples* on the surface of the top-most sediment laid in the basin. The current bedding is due to the turbidity currents generated in the water of the basin. It is important to note that these structures are developed only in the sandstones, and not in the shales or in the argillites. In other words, the current bedding is produced in the coarser grained sediments like those forming the sandstones. At times, the depth of water becomes so much shallow that the top-most layers of the sediments laid in the basin get exposed to the atmosphere. During such periods, the sediments get dried up and produce cracks, which are called as the "mud or sun cracks". Such a feature is often observable in the present-day lakes which get dried up during the summer period. It is also possible that the exposed sediments come under the attack of showers of rain. The impact of such rains creates cavities in the unconsolidated sediments of the basin. These cavities are many times retained in the sediments even after these are converted into hard and compact rocks. This structure is called as "*rain prints*" (Figs. 1.10a, b).

By and large the sedimentary rocks originate in the manner as described above. As a *final stage*, the basin which had received the sediments, it gets uplifted through some geological episode operating below the basin. As a result of such an upliftment, the basin is elevated above the sea level and becomes a part of the landmass. Once fully elevated, such a landmass forms a provenance for the forthcoming cycle of sedimentation. The origin of the sedimentary rocks has been presented in a schematic manner in Fig. 5.1.

Elevation of a part of the crust of the earth producing a provenance.

↓

Physical disintegration and chemical decomposition of the rocks forming the provenance.

↓

Production of soluble and insoluble material from the provenance.

↓

Transportation of the insoluble materials in suspension, and that of soluble ones in solution by the rivers and wind.

↓

Insoluble ferruginous and manganiferous materials, and some silica being left behind at the site of the provenance to form the laterites.

↓

Insoluble materials suffering reduction in size and getting sub-rounded to rounded in shape.

↓

Large sized materials producing pebbles, cobbles, boulders, and getting deposited first in the basin to give rise to the conglomerates.

↓

Finer insoluble materials settling at different rates towards the floor of the basin as layers or beds varying in grain size. This gives rise to the beds or layers of sandstones and shales.

↓

Soluble materials getting precipitated as the carbonates of Ca, Mg, Fe giving rise to limestones and dolomites.

↓

Very fine grained materials acting as cement for the larger grains of the sediments, thus effecting lithification. Authigenic growth around grains further helping to hold the grains more firmly.

↓

Weight of the thick pile of sediments driving out the water held in the pore spaces in between the grains of the sediments.

↓

The sediments undergoing mild degree of metamorphism owing to the slightly increased pressure and elevated temperatures.

↓

Wave currents in the basinal water producing current bedding, ripple marks. Gravity settling of the grains producing graded bedding, and to beds differing in grain size. Occasional exposures of the sediments to the atmosphere producing rain prints and sun cracks.

↓

Upliftment of the basin to form hard and compact rocks.

Fig. 5.1. Schematic representation of origin of the sedimentary rocks.

SUMMARY

1. Sedimentary rocks are formed through joint effort of six stages.
2. Disintegration and decomposition of the crustal rocks, transportation of the sediments, formation of the basin, deposition of sediments in the basin, separation of the sediments into different fractions according to the grain size, compaction of sediments into hard rocks are the 1st, 2nd, 3rd, 4th, 5th and the 6th stage, respectively, of the formation of the sedimentary rocks.
3. Sedimentary beds are formed due to variation of grain size. Banded structures are produced due to variation of colour of the grains. Thickness of beds depends upon the depth of water in the basin, as well as on the quantity of sediments reaching the basin. The graded bedding is developed only in quiet water.
4. Current/false/cross/torrential bedding structures, rain prints, ripple marks are produced in shallow depth of water.
5. Formation of a sedimentary basin and its upliftment are considered to be due to plate tectonics, convection currents generated in sima, intrusion of magmatic body (this gives rise to upliftment of the basin), and trough faulting. Any one or a combination of the geological events, brings about the formation of basin and its subsequent upliftment.

CHAPTER 6

Importance of Sedimentary Rocks

INTRODUCTION

In the previous chapters, the fundamental aspects of the sedimentary rocks have been described. Besides building up of the crust of the earth, some of them are important from the economic and the academic points of view. These aspects have been elaborated in the following paragraphs.

THE ACADEMIC ASPECTS

The geologists often face the problem of establishing the "right side up" position of beds (rocks) that have undergone folding. On such occasions, graded bedding, current bedding and the oscillation ripple marks are made use of, in locating the top surface of the bed. A detailed account of as to how these sedimentary structures are utilised in locating the top and the bottom surfaces of the beds, has been given by Gokhale (1995). Such a knowledge is especially useful to a stratigrapher who is attempting to establish the order of superposition of rocks encountered in the crust of the earth. The above noted primary sedimentary structures are many times retained faithfully, even in the metamorphosed sedimentary rocks (Figs. 1.5a, b, c). Such occurrences are very useful in proving that the present day metamorphic rock was originally a sedimentary and not an igneous one. The inverted or the overturned limb or limbs of folds, and the folds like the recumbent and the fan fold, are often established from the disposition of the above noted primary sedimentary structures, if registered in the folded formations. It is generally regarded by the geologists that a conglomerate bed marks the beginning of the process of sedimentation, and that it marks a surface of unconformity. The conglomerate horizons are made use of (besides other indicators) in establishing the relative ages of the crustal rocks, because each conglomerate horizon (of course, it should be a basal conglomerate bed) marks a new and hence a younger episode of sedimentation. The bottom surface of a conglomerate horizon is designated as a surface of "unconformity" by the stratigraphers.

Fossils are exclusively found in the sedimentary rocks. Palaeontologists are very eager to study the sedimentary rocks with the hope of discovering new fossils. The Standard Stratigraphical Scale is built on the basis of a careful study of fossil bearing sedimentary rocks. Fossils help to understand the organic evolution. Some of the fossils are fantasies like the petrified wood. This is found in sedimentary rocks. Heavy minerals also help to correlate sedimentary rock units and it is called as the "petrographic correlation" (Pettijohn, 1957, p. 130).

Whereas the disposition of the oscillation ripple marks indicates the position of the top and the bottom surfaces of the bedded rocks, the current ripple marks are useful in deriving the direction from which the sediments were brought into the basin. The depth of water in the basin is derived from the occurrence of the ripple marks (current or oscillation ripple marks), sun-mud cracks, current bedding and so on. The presence of these structures indicates a shallow depth of water in the basin. Environment of sedimentation is derived from the size and the shapes analysis of the grains composing the sandstones. This aspect has been elaborated in Chapter 7.

Nature of provenance meaning the rocks that had acted as the source for the sediments, are established by studying the heavy minerals encountered in the sedimentary rocks. The very presence of the sedimentary rocks in any area directly testifies to the fact that in the geological past, a basin had existed at that place. The dispositions of the land and the sea in the geological past, are indicated by the presence of the sedimentary rocks. This constitutes a subject of "Palaeogeography". The palaeoclimate is also established from a careful study of the grains and the pebbles composing the sandstones, and the conglomerates, respectively. Very well rounded sand grains indicate the role of strong wind action. This in turn is indicative of the existence of "desert conditions" in the provenance from which the grains were derived. In fact such well rounded grains are called as the "millet seed sands". Undecomposed laths of felspars, only one face of the pebble being polished and smoothened, these two features together indicate the role of the glaciers, and hence the existence of very cold climate in the region of provenance (Wadia, 1981, p. 172). A combination of unassorted large sized blocks of rocks, and very fine sediments, is again indicative of the existence of the glaciers in the provenance. This gives rise to the beds of "tillite".

ECONOMIC ASPECTS

These are numerous. As has been already noted, the sedimentary rocks are mostly mono-mineralic in composition. As a result of this, some of them are industrially very important. Thus the limestones constitute the important and the main raw material of the cement industry. Its absence will bring the construction of the cement-concrete buildings to a grinding halt. Bauxite, which is an aluminous laterite, is the chief raw material of the aluminium industry. The laterites are always sought after, as a good building material, because these can be easily dressed (cut) when fresh, into any desired shape and size. On exposure to atmosphere, laterite hardens into a hard and resistant rock. Soil is to be considered as a rock, because it is an assemblage of minerals that are produced during the weathering of the crustal rocks. Needless to say that in the absence of the soil covers, there cannot be any agricultural activity at all! Soils therefore should be reckoned as the most important economic sedimentary rocks. Some of the clay rocks sustain the potter industry. Some of the ferruginous shales (ochres) have turned out to be excellent raw material of the paint industry. Most of the world famous iron ore deposits are associated with the banded hematite quartzites, these latter rocks being originally ferruginous-siliceous-sedimentary rocks, that got metamorphosed later into their present form (i.e., banded hematite quartzites). Likewise, the world famous Carara marbles of Italy, the Bhedaghat marbles of Jabalpur, India, and many other occurrences of marbles, were initially limestones which got metamorphosed to the present day marble rocks. The famous Taj Mahal of Agra could not have been built if suitable marble rock (Makrana, Rajasthan) was not formed by the metamorphism of the pre-existing limestones. Some of the shales on metamorphism get converted into slates, and such slates constitute a good paving material.

Some of the beach sands contain radioactive minerals like the pitch-blende, the monazite, the sphene, besides other heavy minerals. Such sands are categorised as the "beach placers". The west coast of Kerala, the east coast of Orissa are good examples of such important placer deposits. In as much as the sandstones are formed from the sands, the beach sands might be considered as the sedimentary rocks, the same way as the soils also are to be reckoned as rocks. Sandstones are the most extensively developed sedimentary rocks. Some of them possess attractive colours, or colour bands. The Red Fort of Delhi is so named because its walls are built of red coloured sandstones. Sandstones could be (relatively) easily cut and trimmed to the desired sizes and shapes. Therefore these rocks yield "dimension stones" like cubes, rectangular slabs, pillars and so on. The grinding stones of the flour mills are made up of suitable sandstones.

The coal seams are associated with the sedimentary rocks alone. The deposits of India are exclusively associated with the rocks of the Gondwana System. It may therefore be said that to look for coal seams, sedimentary rocks of Gondwana age are promising ones. That too sandstones and not shales or the conglomerates, contain the coal seams. Two conglomerate beds of Upper Vindhyan age, are the source rocks of the historically famous Panna and the Golconda diamonds. One horizon occurs between the Kaimur sandstones and the Panna shales. The other horizon is found between the Rewa sandstones and the Ganurgarh shales. In the vicinity of Banaganapalli, very coarse grained sandstones belonging to the Kurnool formations are found to contain diamonds. Thus the conglomerates too are economically useful rocks.

Out of all the sedimentary rocks, the sandstones are the most useful ones. Owing to their remarkable porosity, these form excellent aquifers, both for water and oil. These rocks are therefore most sought after by the geologists in locating sites for bore wells to tap groundwater. A combination of shale at the bottom, and sandstones above it, constitutes a favourable set up for the artesian wells. Such wells are a perennial source of water. Oil and natural gas could be trapped, if the sandstones are folded into anticlines.

SUMMARY

1. Fossiliferous sedimentary rocks immensely help the stratigraphers to correlate different formations and to build up the Standard Stratigraphical Scale. Fauna and flora preserved in these rocks are of immense importance to the zoologists and the botanists.
2. Current bedding, oscillation ripples marks and the graded bedding are made use of, in locating the top-side of the sedimentary beds.
3. Undecomposed laths of felspars, extremely well rounded sand grains (millet seed sands) are of use in deriving the palaeoclimate and building up the palaeogeography of the regions.
4. Sedimentary terranes are excellent groundwater aquifers.

5. Coal seams are associated only with the sedimentary rocks.
6. Conglomerate horizons are likely sources of diamonds.
7. Without limestones, cement industry will come to a grinding halt.
8. Soils are to be considered as important sedimentary rocks.
9. Limestones, shales, banded hematite quartzites, sandstones and aluminous laterites are industrially very useful.

CHAPTER 7

Sedimentary Analysis

INTRODUCTION

In the previous chapters, fundamental information about the sedimentary rocks has been given. Unlike the igneous and the metamorphic rocks, the shape and the size determination of the grains of the sedimentary rocks is absolutely essential. In fact the classification of the sediments into the rudaceous, the arenaceous and the argillaceous categories is based on the actual sizes of the grains. These in turn give rise to three families of the sedimentary rocks. The fourth one, the calcareous group, is not recognised on the basis of the grain size. Therefore, it is absolutely necessary to derive methods of determining the sizes of the grains as accurately as possible. The sediments are deposited under different environments, and these are also required to be established. Such a study is especially taken up by the researchers who are desirous of deriving the evolution of the sedimentary rocks.

It is therefore felt necessary to at least describe the broad features of the methods adopted in the determination of the shapes and the sizes of the grains composing the sedimentary rocks. Such a study is called as the "sedimentary analysis". From the grain size and the shape, the condition of crystallisation of the magma, or that of the recrystallisation in the case of the metamorphic rocks, are determined. In the case of the sedimentary rocks, the sizes and the shapes of the grains, indicate the distance of transportation, as well as the environment of sedimentation. Therefore considerable care and precaution is required to be undertaken while devising the techniques to obtain the actual sizes and the shapes of the grains, as much accurately as possible. The data, thus obtained, is then subjected to the construction of graphs, and calculating values of mean size, sorting, skewness, kurtosis and so on.

CHARACTERISTICS OF SIZE AND SHAPE OF GRAINS IN THE CONTEXT OF SEDIMENTARY ANALYSIS

Prior to proceeding to know about the methods of determining the shape and the size of the grains, it is necessary to elaborate these two parameters. While describing the physical appearance of the grains composing the sedimentary rocks, aspects to be considered are whether the grains are rounded or angular, and their size. The terms "rounded or angular" are considered to connote the shape of the grain. The word rounded, used as an adjective, does not mean that the grain is spherical in shape. As has been noted in Chapter 2, roundness means that the sharp edges of the grains have been made blunt, or have been flattened out. The overall shape of the grain may not be spherical one, but at the same time, sharp edges also will be lacking.

When the size of a grain is to be measured, should all the three dimensions be taken into account, or only one which is the largest? Shape depends upon the habit of the mineral as well as upon its toughness and resistance to the wear and tear taking place during the transportation of the sediments. Neither the original shape nor the size of the minerals composing the source rocks (provenance) are retained during the weathering and transportation processes. The shape as well as the size is a three-dimensional feature, the entire grain is therefore required to be considered. The sedimentologists have therefore thought it appropriate to determine the average diameter of the grain, because the sediments reaching the basin are more or less nearly spherical in shape. It is no doubt an approximation, but at least it helps to devise methods of determining the size and the shape of the grains.

To begin with, the loose sands of the sea coasts and the rivers were studied. Naturally the process of sieving the sands using sieves of different meshes, was adopted. The mesh of a sieve corresponds to the diameter of a spherical body. But the grains of the sedimentary rocks do not possess a spherical shape. Therefore the sieve analysis may not yield the true sizes of the grains. The method of settling rates of sediments could be adopted, because normally the bigger sized material settles down faster than the smaller sized one. It is pertinent to note at this juncture that the pebbles and the cobbles, being bigger in size, settle down faster than the grains of sand. This method however has the drawback of not taking into account the density of the grains. A small sized denser grain will settle down faster than a bigger sized but less dense grain. Likewise, the viscosity of the medium (water of the basin) might adversely affect the settling velocities of the grains.

The sieve analysis or the method of determining the velocities of settling down of the grains is applicable to the loose sands, but it cannot be extended to the hard and compact rocks. Hence the method of thin section analysis is preferred. The shapes of numerous grains are traced with the help of a camera lucida attachment fixed to the microscope. The values of roundness (ρ) are calculated. Since the size of a grain is approximated to a spherical body, roundness may be considered to correspond to the actual shape and size of the grain. In this context, Pettijohn (1957, p. 60) has opined, "Actually in natural sediments, not only are sphericity and roundness closely related with each other, but each in turn is a function of size." Therefore it may be said that the size determination by the thin section method, yields more reliable results.

METHODS OF DETERMINING GRAIN SIZE

As noted above, shape as well as size, is a three-dimensional concept. Therefore all the grains of the rock are needed to be separated from one another, in order to determine the three dimensions of each grain. Separation of grains from one another is called the process of "disaggregation". The grain sizes are determined by (i) the sieve analysis or (ii) by thin section method. Sieve analysis is adopted to the rocks that are rather loosely cemented, and are hence easily disaggregatable. The thin section method is made use of in the cases of hard and compact rocks. These methods are applied only to the sandstones, and not to the shales or the limestones.

(a) The method of sieve analysis

For sieve analysis, the rock specimen has to be disaggregated by gently hammering it, and subjecting the fragments thus obtained, to light crushing in a deep steel mortar, using a rubber pestle. Maximum care has to be taken to retain the original shape and the size of the constituent grains. This is the reason as to why the rock sample needs to be gently hammered, and a rubber pestle to be used, while crushing the fragments of rocks, further, in a steel mortar. The grains thus freed are washed with water using dilute HCl, oxalic acid, hydrogen peroxide and stannous chloride.

Depending upon the grain size of the rock sample (whether coarse, medium or fine), a set of sieves of ASTM standard are used to separate the different size fractions. The sieves are arranged in an ascending order of mesh, the coarsest sieve being placed at the top of the set of sieves. The set of sieves, so arranged, is placed on a "Ro Tap Mechanical Sieve Shaker", and it is subjected to shaking for a period of ten minutes or so. Grains retained on each sieve are collected separately and are weighed. Each sieve has a different mesh value, and it is expressed by the technical term "phi" whose symbol is φ. Phi stands for the diameter of the mesh of the sieve. Values of weight per cent for the grains retained on each sieve, and those of cumulative weight per cent, are calculated. In order to illustrate the result of the sieve analysis, the values obtained by Hegde (1984) have been set out in Table 7.1.

(b) The thin section method

This method is best suited to study the grain size characteristics, because it dispenses with the crushing of the rock sample, and the possibility of adversely affecting the shapes and the sizes of the grains during the crushing operation. However in order to get the true and accurate values, more number of thin sections taken in different directions from each sample, should be studied. The thin section is projected on a micro-projector, and the boundary of each and every grain is traced (Fig. 7.1), taking care that the projection remains orthographic. The values of the diameter of the "inscribing (d_i) and the circumscribing (d_c)" circles for every grain are determined (Fig. 2.1). From such values, the grain size is calculated by using the below noted equation.

$$\text{Size } (\varphi) = \frac{d_i \times d_c}{m}$$

where d_i = diameter of the inscribing circle of the grain,

d_c = diameter of the circumscribing circle of the grain,

m = magnification used while projecting the thin section.

NECESSITY OF PROCESSING GRAIN SIZE DATA

The data obtained from the sieve analysis or from the thin section method, is subjected to the construction of graphs, and statistical calculations, in order to minimise errors creeping in, in either of the methods adopted. In a bed of sandstone, all the grains may not be necessarily of the same size. It then becomes necessary to find out the "frequently encountered grain size". Therefore a resort to mathematical and statistical approach becomes inevitable. In the process, new parameters are also obtained like the kurtosis, the skewness etc. These and the other aspects are described in the forthcoming sections.

CONSTRUCTION OF GRAPHS

The grain size data are processed to obtain class-limits, variation in sizes, if any, and other useful grain size parameters. It is customary to plot the values of the

Table 7.1. Results of sieve analysis (after Hegde, 1984)

Sample number	Mesh number ASTM sieves	Size in φ units	Weight retained in gm	Weight per cent retained	Cumulative weight per cent
202/115	12	–0.75	0.148	0.301	0.301
	14	–0.50	0.340	0.691	0.992
	16	–0.25	0.082	0.167	1.159
	18	0.00	0.538	1.093	2.252
	20	0.25	0.428	0.869	3.121
	25	0.50	2.603	5.288	8.409
	30	0.75	4.968	10.093	18.503
	35	1.00	6.812	13.839	32.341
	40	1.25	18.480	37.544	69.885
	45	1.50	0.126	0.256	70.141
	50	1.75	9.751	19.810	89.951
	60	2.00	1.609	3.269	93.220
	70	2.25	2.047	4.159	97.379
	80	2.50	0.171	0.347	97.726
	100	2.75	0.366	0.743	98.469
	120	3.00	0.358	0.727	99.196
	140	3.25	0.199	0.404	99.600
	170	3.50	0.101	0.205	99.805
	200	3.75	0.068	0.138	99.943
	230	4.00	0.000	0.000	99.943
	240	4.25	0.027	0.055	99.998
	Total		49.222	99.998	
963/322	12	–0.75	0.430	0.860	0.860
	14	–0.50	0.232	0.464	1.324
	16	–0.25	0.273	0.546	1.870
	18	0.00	0.113	0.226	2.096
	20	0.25	0.374	0.748	2.844
	25	—	—	—	—
	30	0.75	0.495	0.990	3.834
	35	1.00	2.704	5.411	9.245
	40	1.25	1.252	2.505	11.750
	45	1.50	—	—	—
	50	1.75	15.508	31.031	42.781
	60	2.00	13.947	27.907	70.688
	70	2.25	4.948	9.901	80.589
	80	2.50	6.134	12.274	92.863
	100	2.75	2.575	5.152	98.015
	120	3.00	0.167	0.334	98.349
	140	3.25	0.408	0.816	99.165
	170	3.50	0.168	0.336	99.501
	200	3.75	0.171	0.342	99.843
	230	4.00	0.045	0.090	99.933
	240	4.25	0.032	0.064	99.997
	Total		49.976	99.997	

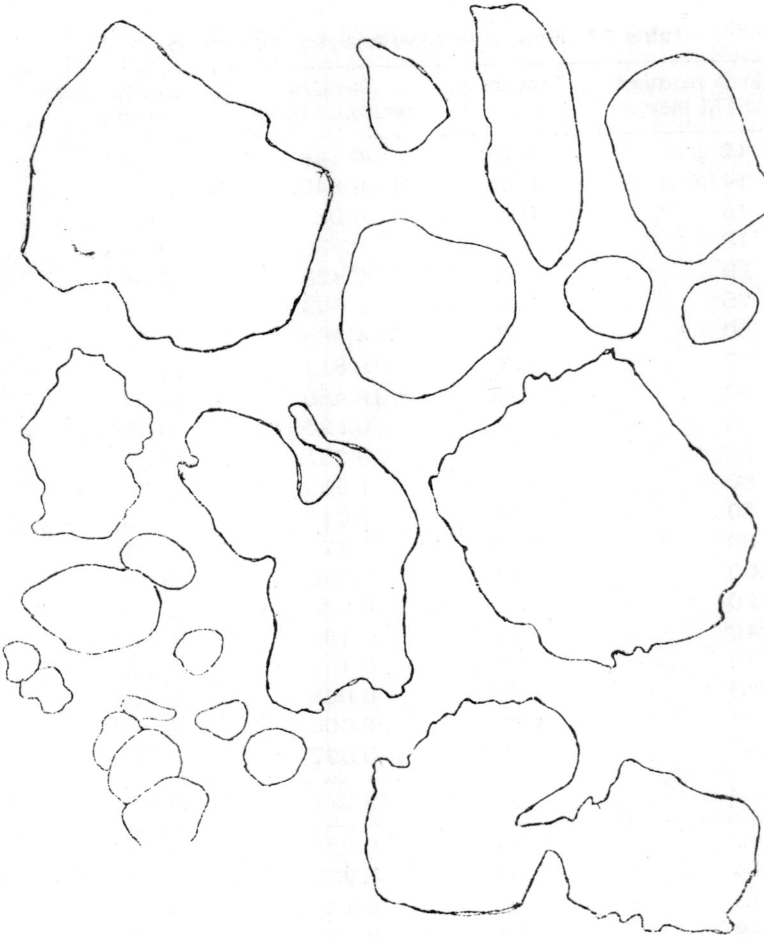

Fig. 7.1. Camera lucida drawings of grains composing quartzarenites, illustrating shapes ×10. *Courtesy :* K. Bhimsen, G.V. Hegde, G.S. Pujar.

cumulative weight per cent against the φ values. Some illustrative graphs have been presented in Fig. 7.2a, b. In the case of data obtained by the thin section method, the cumulative percentage frequency of the grains, is considered, in place of the weight percentage of the grains (Figs. 7.2c, d). It is customary to draw a separate graph for each sample. However it has been observed that the variation in grain size is better revealed if such individual graphs are superimposed on the another, as has been done by Hegde (1984), Pujar (1989) and Bhimsen (1989) (Figs. 7.3a, b). Of course, it is utmost necessary that the rock samples must belong to one bed, and to the same area. Such a superimposition brings out the existence of variation, if any, readily, clearly and more effectively, which might not be readily appreciable on the individual graphs. It is a cumulative approach.

CONSTRUCTION OF FREQUENCY DISTRIBUTION CURVES

It is customary to calculate the average, the arithmetic mean, the geometric mean, the logarithmic mean, and the median values, when a large number of determinations of grain size are available. However all these values are hypothetical ones, in the sense that a rock sample with any such value does not exist, in actuality. It is therefore found it proper to find out the "most frequent value" of grain size, because the probability of encountering samples with such a grain size, is very strong. Determination of grain size for a large number of samples enables to calculate the frequency values for grains belonging to different class-limits (φ values), namely very

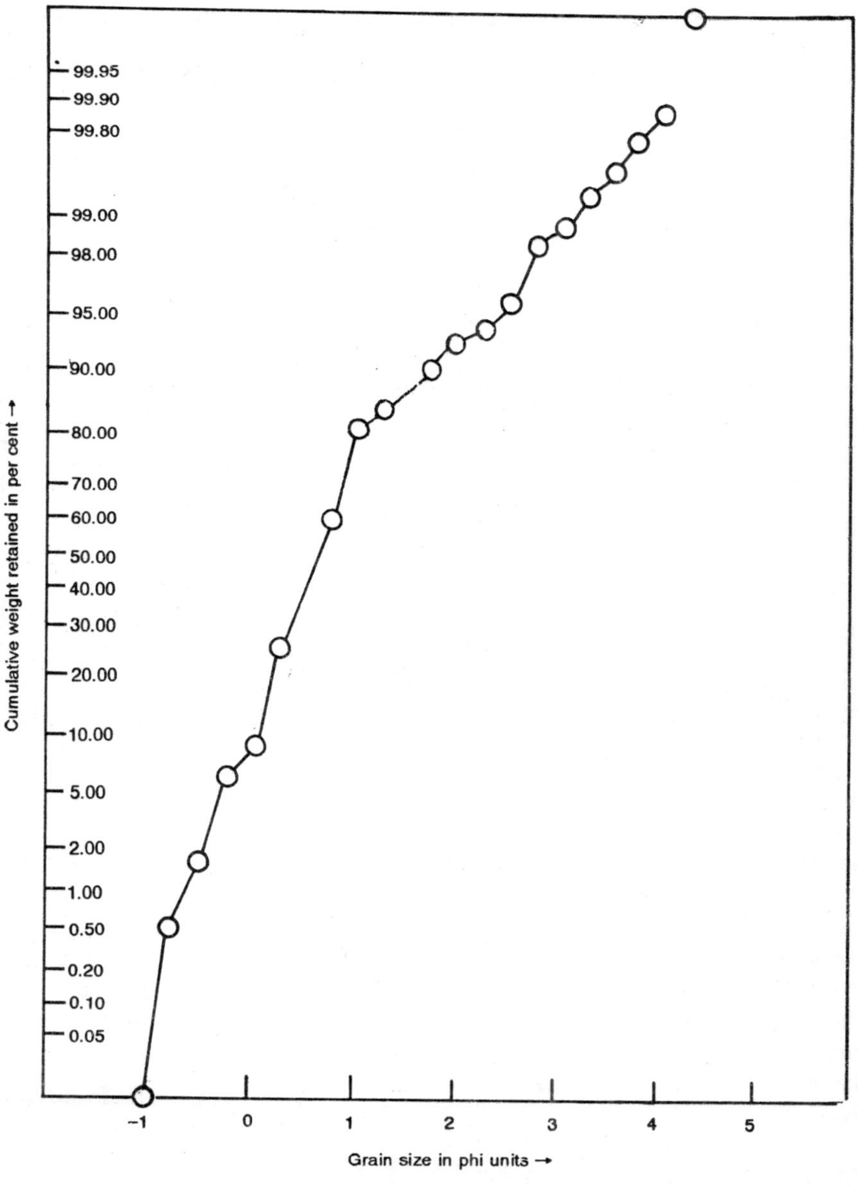

Fig. 7.2a. Cumulative curve drawn on probability per cent ordinate for sample of quartzarenite (Sample No. G.H. 202/115. *Courtesy :* G.V. Hegde, 1984).

fine, fine, coarse and so on. Such frequency values when plotted against the φ values, clearly reveal grain size variation, and its nature. The grain size distribution in rock samples could be (a) uniform or (b) non-uniform. In the non-uniform category, only one size (fine, medium or coarse) might be the most frequent, or two or more sizes might be frequent. The distribution of grain size in rocks is spoken of as the "modal behaviour". If there be only one peak developed in the curve, such rocks are said to possess a "unimodal pattern". If two peaks be developed, the rock is said to possess

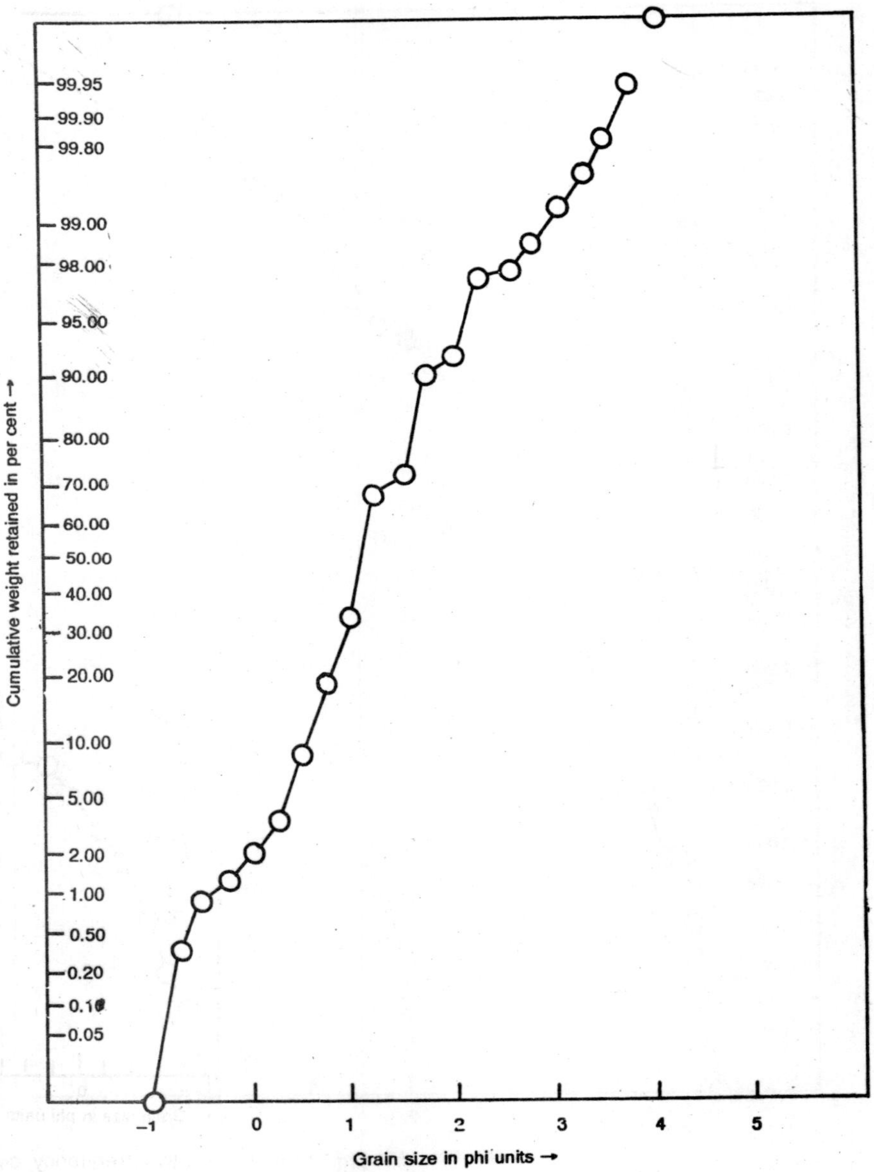

Fig. 7.2b. Cumulative curve drawn on probability per cent ordinate for sample of quartzarenite. (Sample No. GH 260/130. *Courtesy* : G.V. Hegde, 1984).

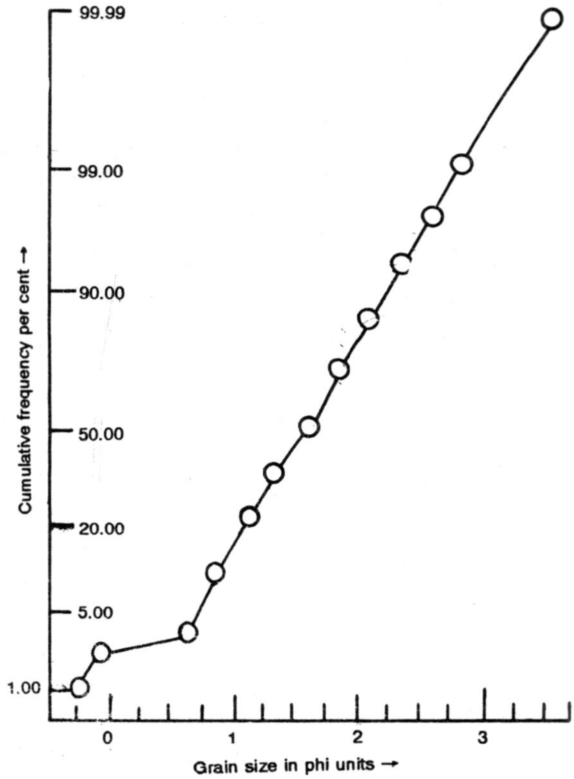

Fig. 7.2c. Cumulative frequency curve drawn for sample of quartzarenite (**Sample** No. PM 70. *Courtesy :* Dr. G.S. Pujar).

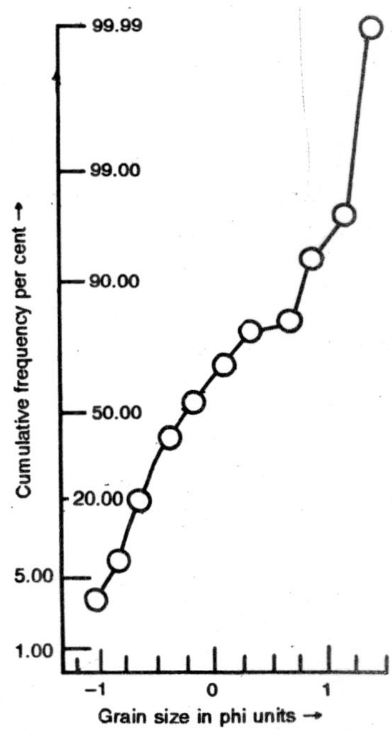

Fig. 7.2d. Cumulative frequency curve drawn for sample of quartzarenite. (Sample No. PM 113. *Courtesy :* Dr. G.S. Pujar).

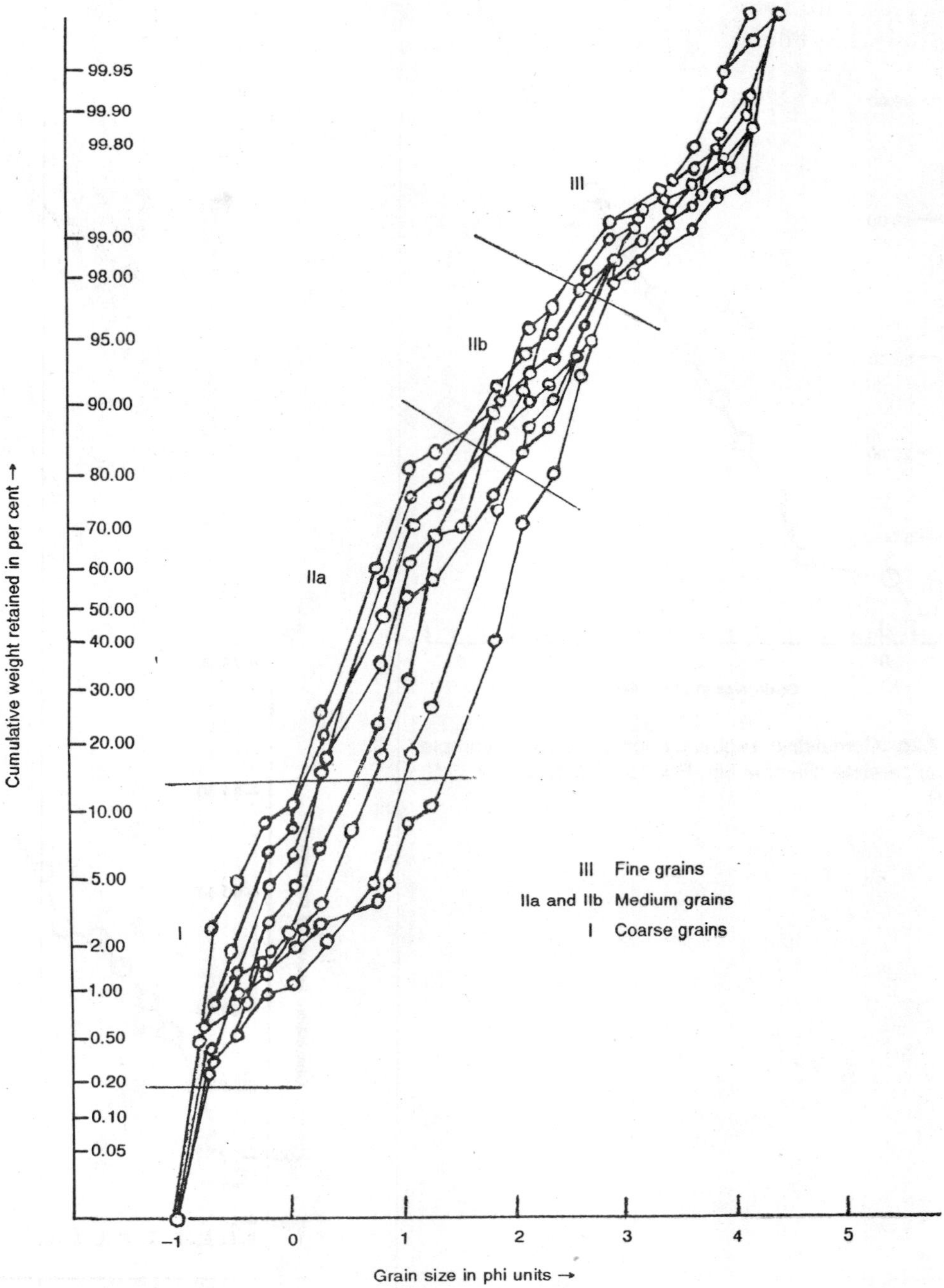

Fig. 7.3a. Superimposition of cumulative curves drawn on probability per cent ordinate for samples of quartz-arenites (after Hegde, G.V. 1984).

a "bimodal pattern". If more than two peaks be developed, then the rock is said to possess a "polymodal pattern". Pujar (1989), and Bhimsen (1989) have drawn such frequency distribution curves, and these have been presented in Figs. 3.5a, b, c, d to illustrate the different modal patterns.

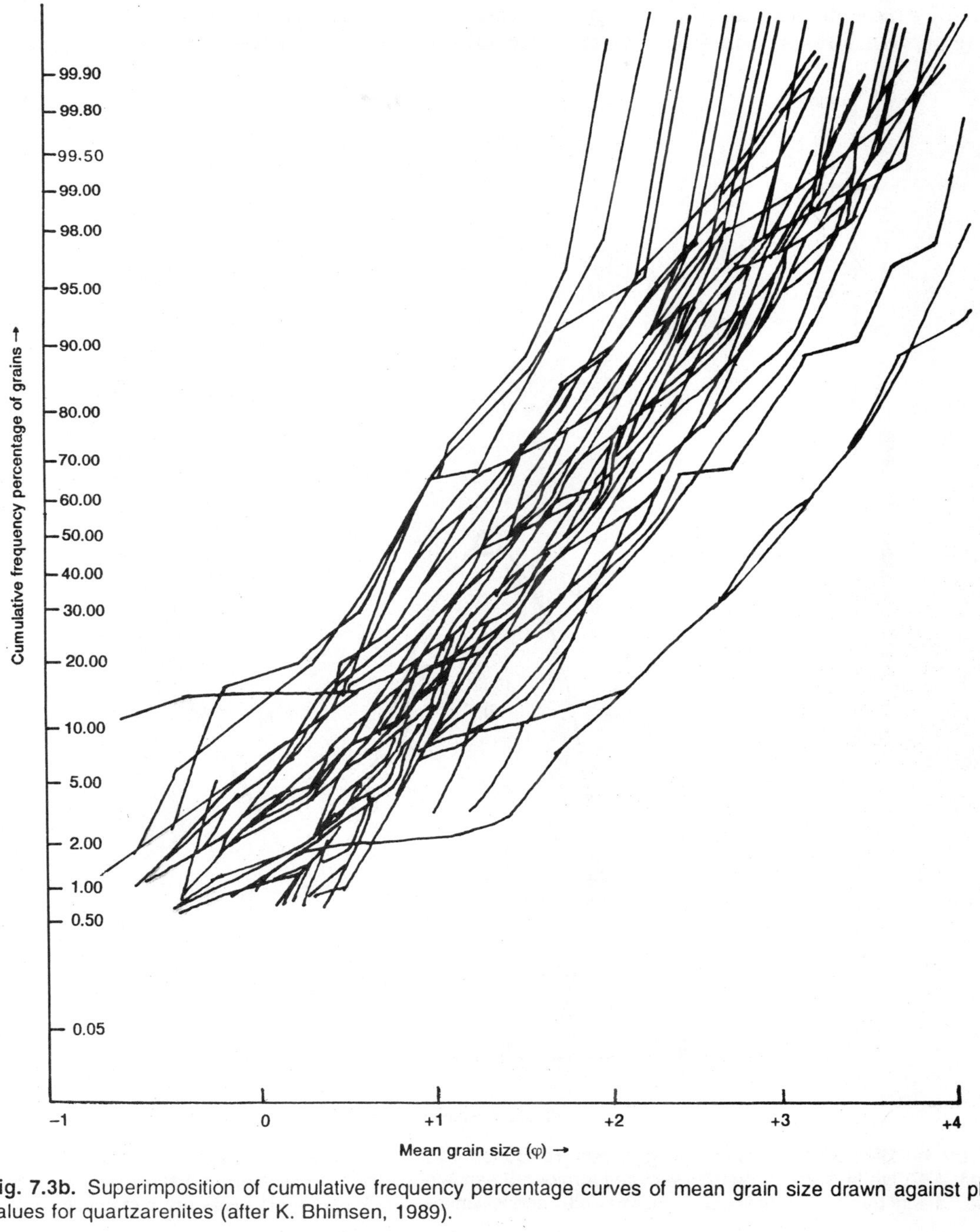

Fig. 7.3b. Superimposition of cumulative frequency percentage curves of mean grain size drawn against phi values for quartzarenites (after K. Bhimsen, 1989).

SIZE DISTRIBUTION STUDIES

Construction of the frequency distribution curves yields different modal patterns which have been described in the previous section. Plotting of cumulative weight per cent (this is obtained from the sieve analysis), or the cumulative per cent of grains (this is obtained from the thin section studies), against the φ values (Figs. 7.2a, b, c, d), also reflects the distribution of grain size in the sample. Data obtained from the sieve analysis, and from the thin section method, are subjected to the "graphical

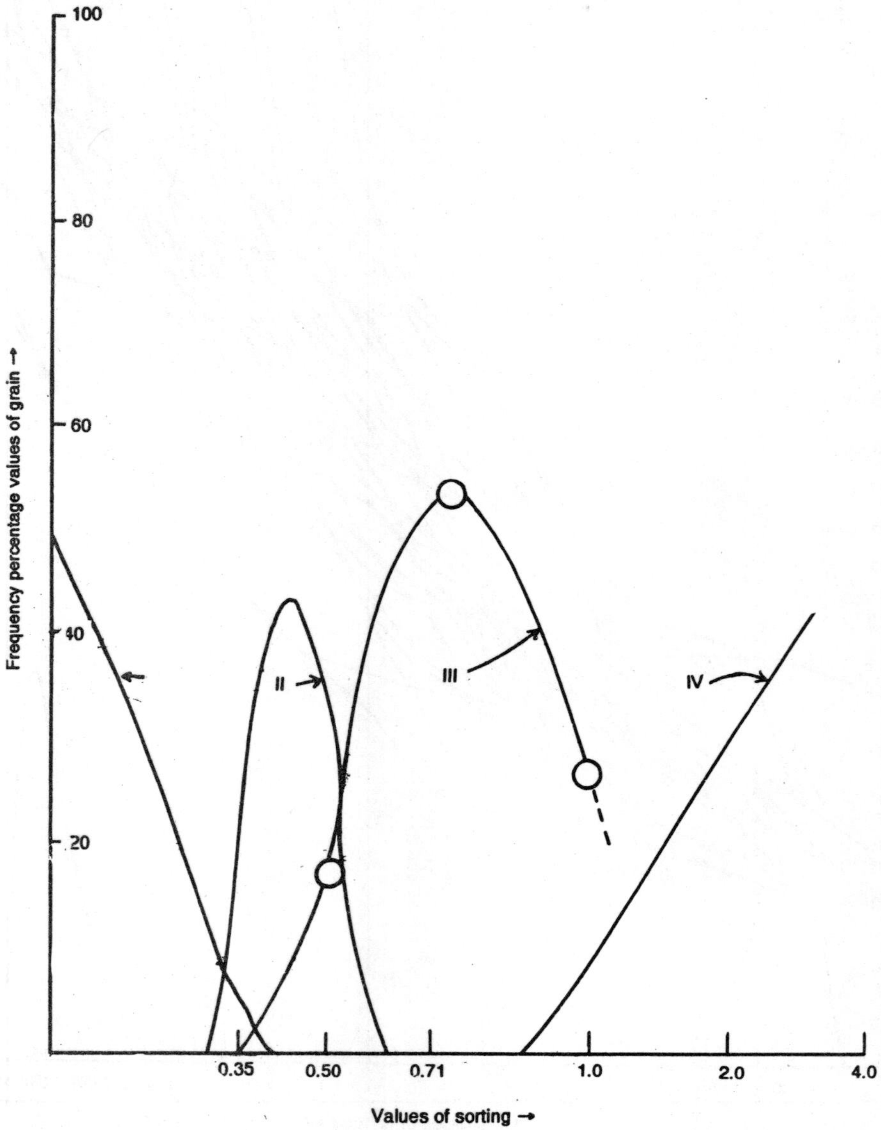

Fig. 7.4a. Different patterns of sorting. I. Very well sorted, II. Well sorted, III. Moderately well sorted, IV. Poorly sorted. Curve III is based on 106 samples (grains) of sandstones from the western and south central parts of the Kaladgi basin. *Courtesy :* G.S. Pujar (1989) and K. Bhimsen (1989). Other curves are hypothetical ones.

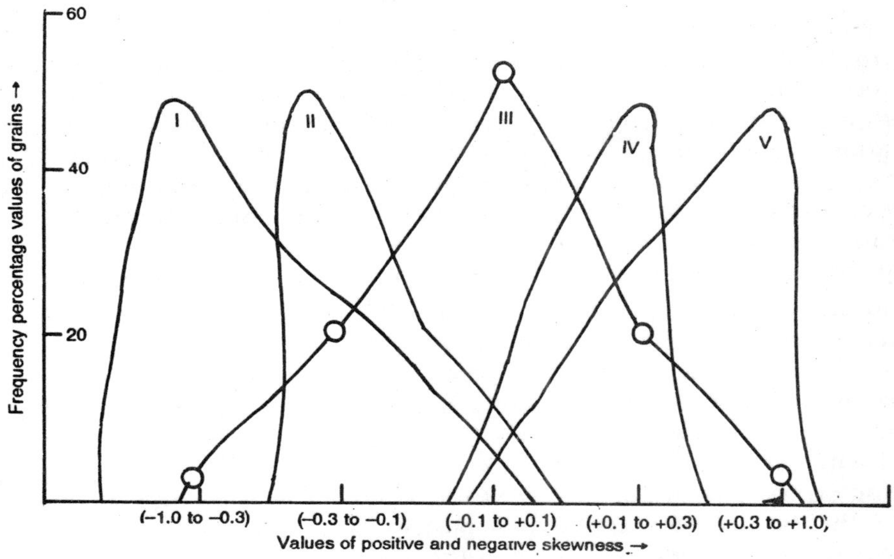

Fig. 7.4b. Different patterns of skewness. I. Very negatively skewed, II. Negatively skewed, III. Nearly symmetrical, IV. Positively skewed, V. Very positively skewed. Curve III is based on data from sandstones of western and southwestern parts of the Kaladgi basin. *Courtesy :* G.S. Pujar (1989) and K. Bhimsen (1989). Other curves are hypothetical ones.

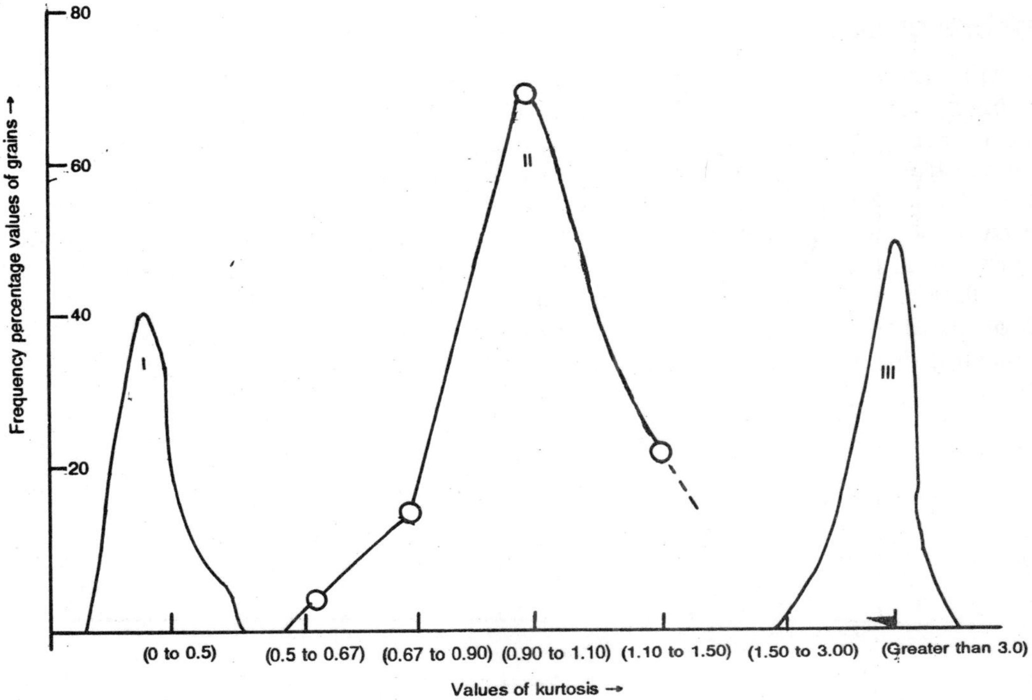

Fig. 7.4c. Different patterns of kurtosis. I. Extremely platykurtic, II. Mesokurtic, III. Extremely leptokurtic. Curve II is based on sandstones from western and southcentral parts of the kaladgi basin. *Courtesy :* G.S. Pujar (1989) and K. Bhimsen (1989). Other curves are hypothetical ones.

and the moment methods", in order to calculate the values of "grain size parameters", namely, mean size, sorting, skewness and kurtosis. A mention has already been made to the term "mean size". In order to clarify the terms "sorting, skewness, and kurtosis", Fig. 7.4a, b, c have been drawn. When the sediments are laid in the basin, a separation (also called as the dispersion or the sorting) takes place according to the sizes of the grains. In some cases, the separation is complete, in some other cases, it is partial, and so on. It therefore becomes necessary to ascertain the degree of separation (sorting) of the grain size. This is called as the process of sorting. From the shape of the frequency curve, the degree of sorting can be appreciated (Fig. 7.4a).

The abscissa of the frequency curve marks the range of grain size, while the values of frequency are marked along the ordinate. The frequency curve can take up any spread over the abscissa. It will be appreciated that the spread of the curve depends upon the actual grain size of the sample. The φ values range from minus to plus, i.e., from fine to coarse, respectively. Therefore the curve may be symmetrically disposed with respect to the zero value of φ. It as well might be asymmetrical, the peak being on the positive or on the negative side of zero φ value. This behaviour is called as the "skewness", and it has positive and negative values. This again is controlled by the frequency values of grain size, whether fine (negative skewness) or coarse (positive skewness). These features have been shown in Fig. 7.4b. Depending upon the sharpness and the curvature of the peak, different categories are recognised. This feature is spoken of as the "kurtosis" of the grains. This has been shown in Fig. 7.4c.

METHODS OF CALCULATING VALUES OF GRAIN SIZE PARAMETERS

The values of mean size, sorting, skewness and kurtosis are calculated from the data obtained from the sieve analysis and/or the thin section method. The data of sieve analysis are subjected to the "graphical" and the "moment" methods. In the graphical method, the φ values corresponding to the 5th, 16th, 25th, 50th, 84th and 95th percentiles are read off from the plottings of the cumulative weight percentage. In the case of the thin section method, the φ values corresponding to the number of grains belonging to the different sizes are read off against the class-limits (Figs. 7.2a, b, c, d). From the percentile values, thus derived, the values of mean diameter ($Md\ \varphi$), mean size (M_Z), sorting (σ_I), skewness (SK_I) and kurtosis (K_G) are calculated by solving the formulae given by Folk and Ward (1957). The formulae are as given below :

$$\text{Mean diameter } (Md\ \varphi) = \varphi_{50},$$

$$\text{Mean size } (M_Z) = \frac{\varphi_{16} + \varphi_{50} + \varphi_{84}}{3},$$

$$\text{Sorting (inclusive of standard deviation) } (\sigma_I) = \frac{\varphi_{84} - \varphi_{16}}{4} + \frac{\varphi_{95} - \varphi_5}{6.6},$$

$$\text{Skewness } (SK_I) = \frac{\varphi_{84} + \varphi_{16} - 2\varphi_{50}}{2(\varphi_{84} - \varphi_{16})} + \frac{\varphi_{95} + \varphi_5 - 2\varphi_{50}}{2(\varphi_{95} - \varphi_5)}$$

$$\text{Kurtosis } (K_G) = \frac{\varphi_{95} - \varphi_5}{2.44(\varphi_{75} - \varphi_{25})}$$

In the "Moment method", the results of the sieve analysis are used directly (meaning not by plotting the data, as done in the case of the "Graphical method"), to obtain the values of mean size, sorting (standard deviation), skewness, kurtosis, and the mean cubed deviation, by solving the below given formulae :

$$\text{Mean size } (\overline{X}_\varphi) = \frac{\sum fm}{n},$$

$$\text{Standard deviation (sorting) } (\sigma_\varphi) = \sqrt{\frac{\sum f(m - \overline{X}_\varphi)^2}{100}},$$

$$\text{Skewness } (SK_\varphi) = \frac{\sum f(m - \overline{X}_\varphi)^3}{100 \times \sigma\varphi^3},$$

$$\text{Kurtosis } (K_\varphi) = \frac{\sum f(m - \overline{X}_\varphi)^4}{100 \times \sigma\varphi^4},$$

$$\text{Mean cubed deviation } (\alpha_3\sigma^3) = \frac{1}{100} \times f(m - \overline{X}_\varphi)^3,$$

where f = weight per cent (frequency) of each grain size grade,

m = mid point of each grain size grade in φ values,

n = total number of samples which is 100 when f is in per cent.

The φ moment statistics is the equivalent of the graphical statistics proposed by Folk and Ward (op. cit.), i.e.,

φ moment statistics	**Graphical statistics**
Mean size ($\overline{X}_\varphi$)	Mean size (M_Z)
Standard deviation (sorting) (σ_φ)	Sorting (σ_I)
Skewness $\left(\frac{SK_\varphi}{2}\right)$	Skewness (SK_I)
Kurtosis $\left(\frac{K_\varphi}{3}\right)$	Kurtosis (K_G)

It is pertinent to note that the values of the above described parameters obtained by the Graphical and the Moment methods are nearly identical.

CLASSIFICATION OF GRAIN SIZE PARAMETERS

Folk and Ward (op. cit.) have classified each grain size parameter into subdivisions. It is to be remembered that the grains vary from coarse to fine. Therefore it will be necessary to stipulate actual values corresponding to the different sizes. Likewise, the other parameters, namely the sorting, the skewness etc., also have been subdivided by Folk and Ward (op. cit.). In Table 7.2, the range values of the different size parameters have been set out.

The subdivisions within each grain size parameter (Table 7.2) are of immense use in deducing the conditions of sedimentation, the distance of transport of the sediments, the agent of transportation of the sediments, and so on. Subdivisions within the "mean size" help to recognise varieties like "coarse, medium and fine" grained sandstones, more precisely. On the basis of "sorting", categories like "poorly, moderately etc.," sorted sandstones can be distinguished. The parameter of "skewness" helps to distinguish varieties within the sandstones like "very negatively,

Table 7.2. Range values of size parameters

Size parameter	Standard range values and their class terms as given by Folk and Ward (op. cit.)	
	Range values	Class terms
Mean size (M_Z)	-1.00 to 0.00 0.00 to 1.00 1.00 to 2.00 2.00 to 3.00 3.00 to 4.00	Very coarse sand Coarse sand Medium sand Fine sand Very fine sand
Sorting (σ_I)	Less than 0.35 0.35 to 0.50 0.50 to 1.00 1.00 to 2.00 2.00 to 4.00	Very well sorted Well sorted Moderately sorted Poorly sorted Very poorly sorted
Skewness (SK_I)	-1.00 to -0.30 -0.30 to -0.10 -0.10 to 0.10 0.10 to 0.30 0.30 to 1.00	Very negatively skewed Negatively skewed Nearly symmetrical Positively skewed Very positively skewed
Kurtosis (K_G)	Less than 0.50 0.50 to 0.67 0.67 to 0.90 0.90 to 1.11 1.11 to 1.50 1.50 to 3.00 Greater than 3.00	Extremely platykurtic Very platykurtic Platykurtic Mesokurtic Leptokurtic Very leptokurtic Extremely leptokurtic

nearly symmetrical etc.," types. On the basis of "kurtosis", varieties like "extremely platykurtic, mesokurtic etc.," sandstones can be identified. Such a distinction is not possible merely by the microscopic examination of the grains.

DETERMINATION OF ENVIRONMENTS OF SEDIMENTATION

The grain size parameters are further useful in reckoning (deducing) the environments of sedimentation. The wind, the running water, nearly share the entire "process of sedimentation". In other words it means that there are two main environments of deposition namely, the aeolian (wind) and the basin (meaning water bodies like lakes, seas, oceans). The water in the basin could be shallow, deep, marine, fresh etc. In order to deduce these varieties of environments, Friedman (1961, 1967) has suggested graphs, while Sahu (1964) has put forth equations. Friedman (1961) makes distinction between the "Beach and the River environments" with the help of graphical plotting. In one of them, the boundary between the Dune (wind) and the "River and Beach" environments, has been drawn. The plot (graph) of values of mean size against those of the "inclusive standard deviation" (Fig. 7.5), separates the Dune, the River and the Beach environments from one another. The plots of

(i) Mean Size against Standard Deviation (Fig. 7.6),
(ii) Mean Cubed Deviation against Standard Deviation (Fig. 7.7),
(iii) Skewness against Standard Deviation (Fig. 7.8), and
(iv) Skewness against Inclusive Standard Deviation (Fig. 7.9),

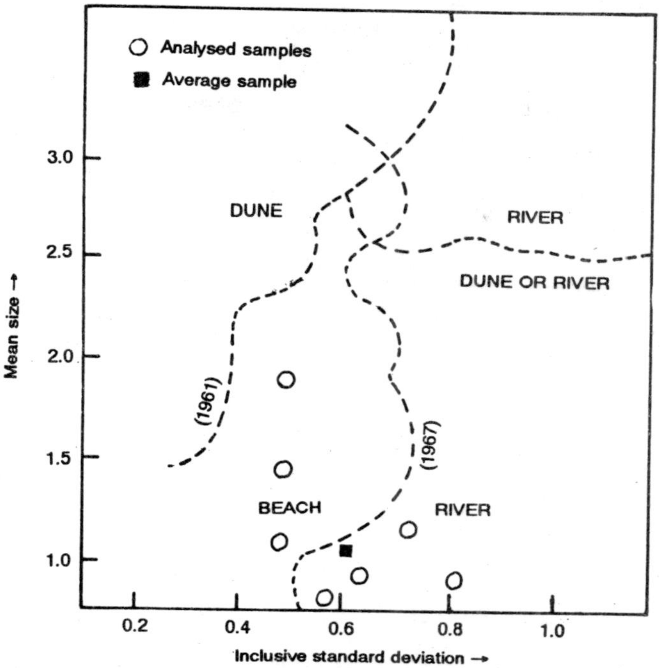

Fig. 7.5. A plot of mean size against inclusive standard deviation. The boundary lines (dashed) are after Friedman (1961 and 1967) (after Hegde, G.V. (1984).

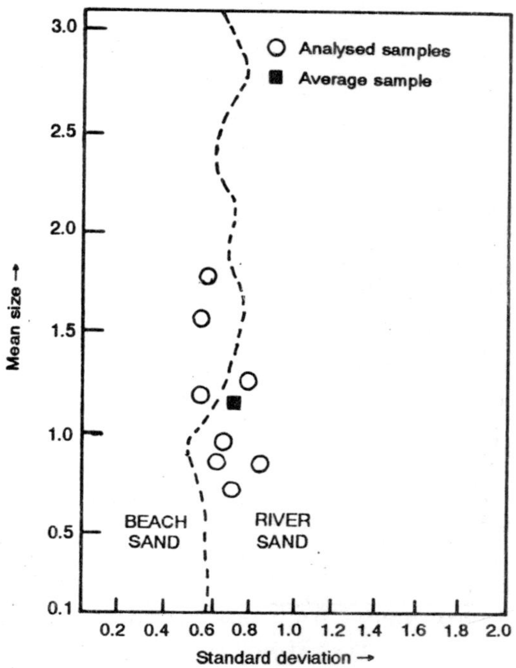

Fig. 7.6. A plot of mean size against standard deviation. The boundary line (dashed) is after Friedman (1967) (after Hegde, G.V. 1984).

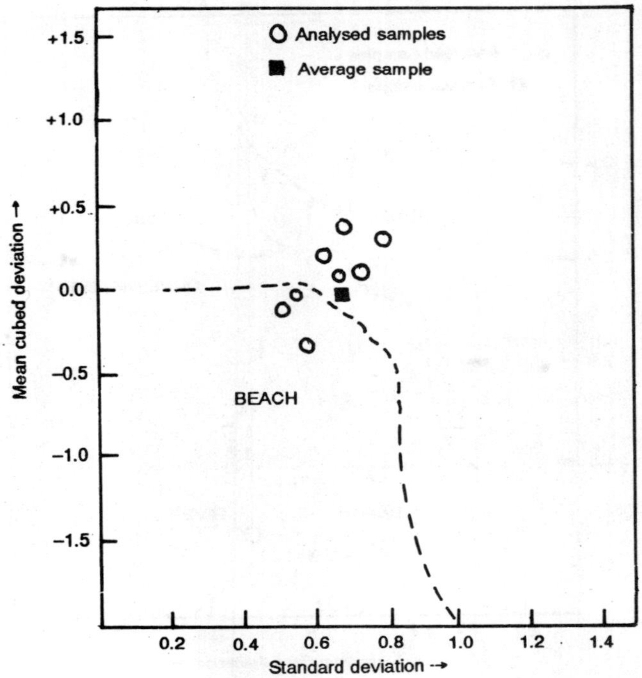

Fig. 7.7. A plot of mean cubed deviation against standard deviation. The boundary line (dashed) is after Friedman (1967) (after Hegde, G.V. 1984).

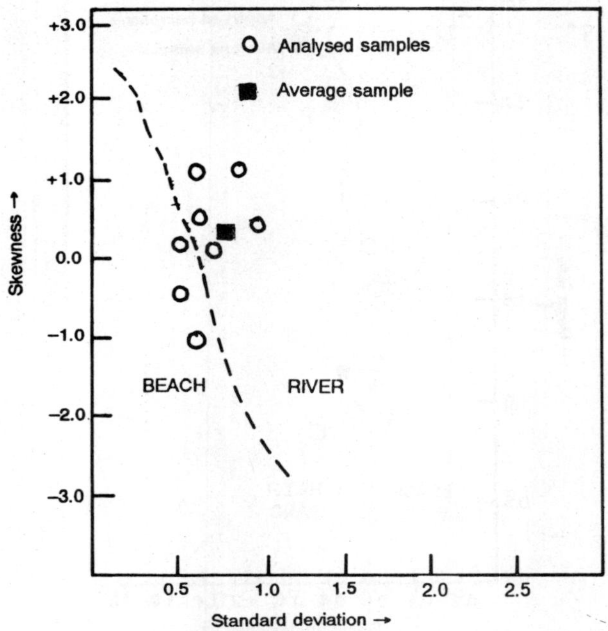

Fig. 7.8. A plot of skewness against standard deviation. The boundary line (dashed) is after Friedman (1967) (after Hegde, G.V., 1984).

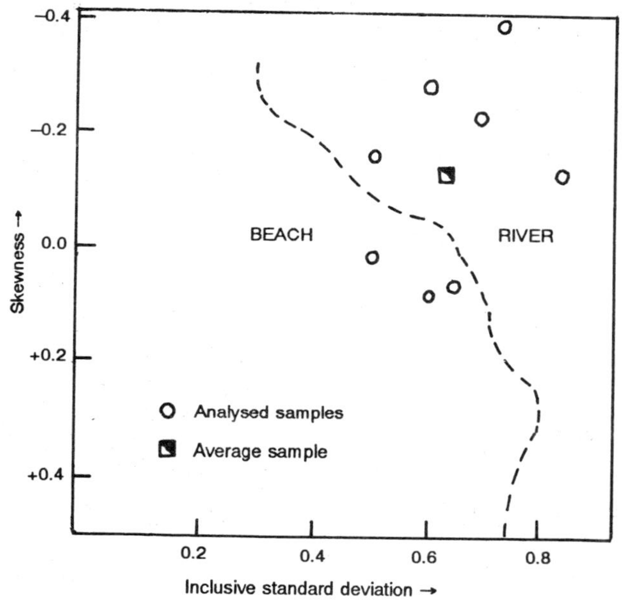

Fig. 7.9. A plot of skewness against inclusive standard deviation. The boundary line (dashed) is after Friedman (1967) (after Hegde, G.V., 1984).

demarcate the Beach and the River environments. These are all binary plots. Hedge (1984) has suggested a trivariate diagram based on the values of skewness, mean size, and sorting. He has even considered the positive and the negative values of skewness, separately. Further, whereas Friedman's (1961) binary plots are based on studies of the loose sands, the trivariate plot of Hegde (op. cit.), is devised on the hard and compact rocks. As such it is more useful to the geologists, because their studies are on hard rocks, and not on the loose sands. In Fig. 7.10, Hegde's trivariate diagram has been presented. Pujar (1989) and Bhimsen (1989) have made use of the trivariate diagram of Hegde (op. cit.) successfully. Gokhale et al. (1989) have demarcated the fields of River and Beach, more precisely (Fig. 7.11).

The environment of sedimentation is deduced by adopting the "Discriminant Functions" proposed by Sahu (1964). He recognises five environments namely, (i) Aeolian, (ii) Beach, (iii) Shallow marine, (iv) Fluvial (Deltaic) and, (v) Turbidity. As a first step, the values of the grain size parameters (mean size, sorting, skewness and kurtosis) are obtained by adopting the methods of sieve analysis, or the thin section method. The values thus obtained are then made use of in solving the below noted four equations of Sahu (1964) :

1. $Y_{Aeolian:\,Beach} = -3.5688M_Z + 3.7016\sigma_I^2 - 2.0766SK_I + 3.1135K_G$

   $\overline{Y}_u$ = less than –2.7411 Beach Environment,

   $\overline{Y}_u$ = greater than –2.7411 Beach Environment.

2. $Y_{Beach:\,Shallow\ Marine} = 15.6534M_Z + 65.7091\sigma_I^2 + 18.1071SK_I + 18.5043K_G$

   $\overline{Y}_u$ = less than 65.3650 Beach Deposition,

   $\overline{Y}_u$ = greater than 65.3650 Shallow Marine Deposition

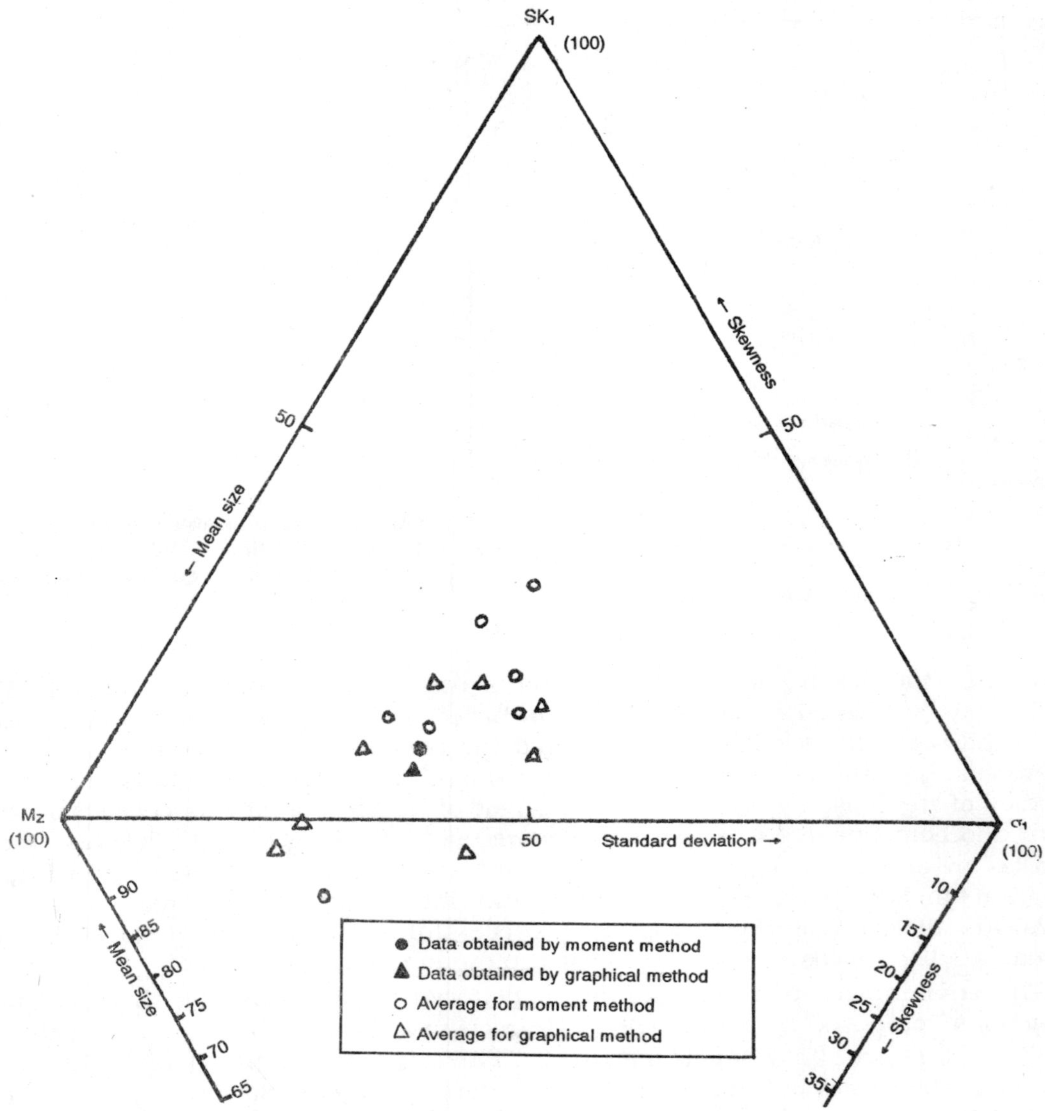

Fig. 7.10. Triangular variation diagram of +ve and –ve skewness, mean size and standard deviation for samples of quartzarenites (after Fig. 17 of Hegde, G.V. 1984).

3. $Y_{Shallow\ Marine:\ Fluvial} = 0.285M_Z - 8.7604\sigma_I^2 - 4.8932SK_I + 0.0482K_G$

$\overline{Y}_u$ = less than –7.4190	Fluvial (Deltaic) Deposition
$\overline{Y}_u$ = greater than –7.4190	Shallow Marine Deposition

4. $Y_{Fluvial:\ Turbidity} = 0.7215M_Z - 0.4030\sigma_I^2 + 6.7322SK_I + 5.2927K_G$

$\overline{Y}_u$ = less than 9.8433	Turbidity Current Deposition,
$\overline{Y}_u$ = greater than 9.8433	Fluvial (Deltaic) Deposition.

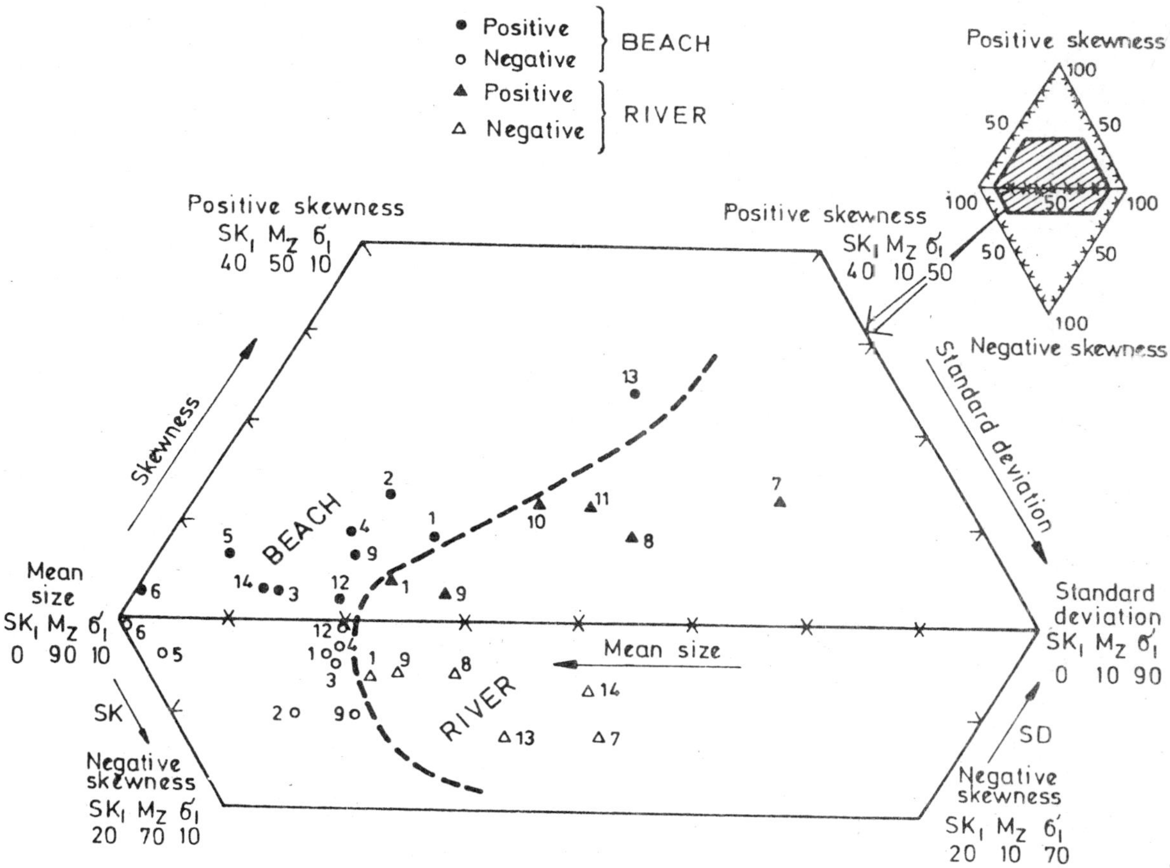

Fig. 7.11. Fields of beach and river environments demarcated.

On solving the above given four equations, the environment of sedimentation can be derived. Equation 1 makes the distinction between the Aeolian and the Beach environments. Distinctions within the Beach environment (whether shallow marine, fluvial, deltaic or turbidity) are made from the values obtained from equations 2, 3 and 4. Thus from the foregoing account, it is quite evident that the sedimentary analysis is very useful in the sedimentological studies, because it yields more and valuable information which could have otherwise gone unnoticed. The sedimentary analysis is especially indispensable to the researchers dealing with the evolution of the sedimentary rocks.

SUMMARY

Below noted features are noteworthy :

1. Grain size is determined by sieve analysis or by the thin section method.
2. Sieve analysis is applied to the loosely cemented rocks and to the loose sands. Thin section method is adopted to hard, compact rocks, and is hence more useful to the geologists.
3. Only sandstones are subjected to the sedimentary analysis.

4. Shape of a grain is generally considered to be rounded one. With the help of camera lucida drawings of the shapes of the grains, the degree of roundness is ascertained.
5. Data obtained from the sieve analysis, or from the thin section method, are utilised to construct graphs, and frequency distribution curves.
6. From the values of grain size, the values of the "grain size parameters" namely mean size, sorting, skewness and kurtosis, are calculated.
7. Adopting the equations of Sahu, the environments of sedimentation are derived from the values of grain size parameters.
8. A trivariate diagram based on the values of sorting, positive and negative skewness, and the mean size, has been suggested to demarcate the fields of River and the Beach environments. It is adoptable to hard rocks, and is hence very useful to the geologists.

Bibliography

Abhinaba Roy and S.K. Biswas (1982). A note on the "volcanic conglomerate"/ pyroclast horizon near Palkanmaradi in the Hutti Schist belt, *Current Science*, Vol. 23, Oct. 1982, pp. 506-510.

Bhimsen, K. (1989). Geological studies of the rocks of Ajra-Gandhinglaj area, Kolhapur district, Maharashtra, India. Ph.D. thesis, Karnatak University, Dharwad.

Chandrasekhara Gowda, M.J. (1976). Geology of the area around Kaladgi, Bijapur district, Karnataka. Ph.D. thesis, Mysore University, Mysore.

Dake, C.L. and Brown, J.S. (1925). **Interpretation of topographical and geologic maps with special reference to determination of structure.** 1st edition, McGraw Hill Co. Inc., New York and London, pp. 126-204.

Desai, H.D. (1991). Geological studies at and around Jambunath hills, Bellary district, Karnataka State, Ph.D. thesis, Karnatak University, Dharwad.

Folk, R.L. and Ward, W.C. (1957). Brazos river bar—a study in the significance of grain size parameters. *Jour. Sed. Petr.* Vol. 27 No. 1, pp. 3-26.

Folk, R.L. (1968). **Petrology of sedimentary rocks** (Revised edition), Austin, Texas, Hemphills Book Store.

Friedman, G.M. (1961). Distinction between dune, beach and river sands from their textural characteristics. *Jour. Sed. Petr.* Vol. 31, pp. 514-529.

Friedman, G.M. (1967). Dynamic processes and statistical parameters compared for size frequency distribution of beach and river sands. *Jour. Sed. Petr.* Vol. 37, pp. 327-354.

Gokhale, N.W., Bhimsen, K. (1989). Texture, mineralogy, sedimentology and depositional environment of sand grains of quartzarenitic rocks of Ajra basin, Kaladgi formation, Kolhapur district, Maharashtra. *Kar. Univ. Jour. Sc.* XXXIV, pp. 49-64.

Gokhale, N.W., Bhimsen, K., Chavadi, V.C., Hegde, G.V. and Pujar, G.S. (1989). A trivariate distinction between River and Beach environments of ancient sedimentary rocks—a new approach. *Acta Geologica Hungarica*, Vol. 32/3-4, pp. 349-357.

Gothe, N.N. (1973). Geology of the granites and the associated rocks of Mundargi-Hadagalli area. Ph.D. thesis, Kar. Univ., Dharwad.

Govind Rajulu, B.V. and Chandrasekhara Gowda, M.J. (1966b). Stromatolytic lime-stones from the Kaladgi formations around Lokapur, Bijapur district, Mysore State. *Jour. Mys. Univ.* Vol. 20, pp. 7-16.

Govind Rajulu, B.V. and Chandrasekhara Gowda, M.J. (1968a). Authigenic plagio-felspar from the Upper Kaladgi quartzites of Lokapur, Bijapur district, Mysore State, India. *Bull. Geol. Soc. Ind.* Vol. 5, No. 4, pp. 124-127.

Govind Rajulu, B.V. and Chandrasekhara Gowda, M.J. (1968b). Algal stromatolites from the southwestern part of the Kaladgi basin, Lokapur, Mysore State, India. *Jour. Sed. Petr.* Vol. 38 (A), pp. 1059-1064.

Govind Rajulu, B.V. and Nagraj, H.R. (1969). Authigenic tourmalines from Lower Kaladgi sandstones, Jamkhandi, Mysore State. *Jour. Sed. Petr.* Vol. 39, pp. 391-394.

Hegde, G.V. (1984). Structural and sedimentological studies of the rocks of Mudakavi-Lakhamapur area, Belgaum-Bijapur districts, Karnataka State. Ph.D. thesis, Kar. Univ., Dharwad.

Holmes, A. (1987). **Principles of physical geology.** ELBS publication, 3rd edition.

Jayaprakash, A.V., Sundaram, V., Hans, S.K. and Mishra, R.N. (1987). Geology of the Kaladgi and Badami basins Karnataka. *Geol. Soc. Ind. Mem.* 6, pp. 201-225.

Koppad, V.B. (1976). Geology of the area around Belhatti and Bannikoppa, Shirhatti taluka, Dharwad district, Ph.D. thesis, Kar. Univ., Dharwad.

Kouhsari, A.H. (1986). Geological studies of bauxite deposits of Shrivardhan area, Kolaba district, Maharashtra State, India. Ph.D. thesis, Kar. Univ., Dharwad.

Krishnan, M.S. (1968). **Geology of India and Burma.** Higginbothams (P) Ltd., Madras.

Krumbein, W.C. and Pettijohn, F.J. (1938). **Manual of sedimentary petrology. The Century Earth Science Series.** D. Appleton-Century Company Inc., New York.

Maclaren, J.M. (1906). Notes on auriferous tracts in southern India. *Rec. Geol. Surv. Ind.* Vol. XXXIV, pp. 90-105.

Müller, G. (1967). **Methods in sedimentary petrology, Pt. I.** Hufner Publishing Co., New York/London.

Pettijohn, F.J. (1957). **Sedimentary rocks**, 2nd edition, Harper Geoscience Series. Oxford Books and Stationery Co., Calcutta.

Pettijohn, F.J. (1984). **Sedimentary rocks**, 3rd edition, CBS Publishers & Distributors, Delhi.

Pujar, G.S. (1989). Geology of the area east of Manoli, Belgaum district, Karnataka state. Ph.D. thesis, Kar. Univ., Dharwad.

Puranik, S.C. (1979). Iron formations of Nagavi-Doni area, Dharwad district, Karnataka State. Ph.D. thesis, Kar. Univ., Dharwad.

Ravindra, B.M. and Ranganathan, N. (1994). **GeoKarnataka** (Edited), Mysore Geological Department Centenary volume brought out by the Karnataka Assistant Geologists Association, Department of Mines and Geology, Bangalore.

Sahu, B.K. (1964a). Depositional mechanism from the size analysis of clastic sediments. *Jour. Sed. Petr.* Vol. 34, pp. 73-83.

Sahu, B.K. (1964b). Significance of the size distribution statistics in interpretation of depositional environment. *Res. Bull. Punjab Univ.* (N.S.) Vol. 15, pt. 3-4, pp. 213-219.

Sahu, B.K. (1965a). Transformation formula. *Jour. Sed. Petr.* 35, pp. 969-972.

Sahu, B.K. (1982a). Multigroup discrimination of river, beach and dune sands using roundness statistics. *Jour. Sed. Petr.* Vol. 52, pp. 779-784.

Sahu, B.K. (1983). Multigroup discrimination of depositional environments using size distribution statistics. *Ind. Jour. Earth Sc.* 10, pp. 20-29.

S. Omara, Wagih, W., Bishara and M. Nasr (1974). Grain size parameters and palaeoenvironments of the Nubia sandstones. *Jour. Sed. Petr.* Vol. 44, pp. 136-144.

Stutton, L., Basu, A. and Mack, G.H. 1981. Climate and origin of quartzarenites. *Jour. Sed. Petr.* Vol. 51, No. 4, pp. 1235-1246.

Venkatachala, B.S. and Maheshwari, H.K. (1991). Indian Gondwana, (Edited), *Geol. Soc. Ind. M.* 21.

Visher, G.S. (1969). Grain size distributions and depositional processes. *Jour. Sed. Petr.* 39 (3), pp. 1074-1106.

Viswanathiah, M.N. (1977). Seminar volume on Kaladgi, Badami and Cuddapah sediments (Edited). *Ind. Mineralogist.* Vol. 18, pp. 1-122.

Wadia, D.N. (1981). **Geology of India**, 4th Edition. Tata McGraw Hill Publishing Co. Ltd., New Delhi.

Wentworth, C.K. (1922). A scale of grade and class terms for clastic sediments. *Jour. Geol.* Vol. 30, pp. 377-392.